A Synthesis of Quantum Gravity

First Edition

Dr. Robert Nieves

Library of Congress Control Number: 2021903524

ISBN: 9798715826565

A Synthesis of Quantum Gravity

This book is ideal for students, researchers, and readers in all areas of cosmology, quantum mechanics, special and general relativity. There are three main parts of the book focusing on waves, energy, and quantum mechanics, to discuss the greatest questions to physical reality. The synthesis explains in detail the quantum gravity aspects that serve as a foundation to the Special and the General Theory of Relativity. Quotes are provided since the good wisdom in the message of a quote can make a lasting impact on the motivation, inspiration, and well-being of the reader.

There are three main messages in this book. The first is that the spatiotemporal wave function is the fundamental centerpiece of physics. The second message is that it is all a matter of waves of probability, including time. Quantum Mechanics and General Relativity are wave theories of physical reality. The third message is that both space and time are nonlinear.

Is General Relativity emergent from the quantum scale? How does General Relativity emerge from a Quantum Theory like A Dynamic Theory of Space-Time? The principle of quantum general relativity is inherent in A Dynamic Theory of Space-Time as a theory of quantum gravity. This book explains and describes why the wave is transcendental and relativistic in its nature. The attributes of the General Theory of Relativity spring from a Quantum Theory of Quantum Mechanics. The author describes in detail how the quantum relativistic attributes of the spatiotemporal wave function scales up to the classical world arounds us in our universe.

Robert Nieves has a diversified professional experience in engineering, teaching, international business administration, and physics and cosmology research. Dr. Nieves holds a Bachelor of Science in Electrical Engineering from the Illinois Institute of Technology and an MBA and a DIBA from Nova Southeastern University in Florida.

Dedicated to my parents and family

CONTENTS

PART I

WAVES

5. Is a dynamic block universe deterministic or indeterministic?

6. Is space-time continuous or discrete?

7. The information paradox for a black hole.

PART II

ENERGY

Pressure equals Energy Density: A Natural Law

1. What is a field of force?

2. The photonic field.

3. Why is acceleration represented by a unit of spatial distance per every two units of temporal distance?

4. Could spatiotemporal curvature be represented as probability? How can the Einstein field equations be represented by a wavefunction?

5. What is the principle of equivalence between pressure and energy density? How is the wave function related to the pressure of spatiotemporal divergence?

PART III

QUANTUM MECHANICS

Time and Space are Nonlinear

1. How does the human mind measure time? Is the human mind using more than one clock?

2. The light cone.

3. Natural spatiotemporal waves.

Quotes by Category

1. Physics

2. Science

3. Mathematics and Geometry

4. Motivational and Inspirational

5. Entertaining

PART I

WAVES

Chapter 1

The Wavefunction of Probability

§ 1. What are the advanced and retarded waves at a spatiotemporal point and their directions?

The advanced and retarded waves are spatiotemporal waves whose directions are opposite. In an expanding six-dimensional universe, every spatial dimension has a conjugate temporal dimension, and the spatial resultant wave is opposite to the temporal resultant wave. In terms of the resultant waves, the advanced wave travels towards the past and the retarded wave travels towards the future. Hence, space-time may contract, expand, or stay the same, depending on the amplitude of each advanced or retarded wave, as an attribute of space-time at an arbitrary point. (Baker, 1987)

In an expanding region of the universe, the retarded wave has a larger amplitude than its advanced wave. An advanced wave and a retarded wave exist simultaneously at any spatiotemporal point in any of the six Cartesian coordinate directions of six-dimensional space-time or in any of the directions of the resultant waves.

The retarded wave may be represented by a "ket", $|\Psi^+\rangle$, with complex spatial coefficients of the spatial basis vectors, and the advanced wave may be represented by a "bra", $\langle\Psi^-|$, with complex conjugate temporal coefficients of the temporal basis vectors. At an arbitrary spatiotemporal point, we may represent the advanced and retarded waves in bra-ket notation, $\langle\Psi^-|\Psi^+\rangle$, as shown below in six-dimensional space-time. Therefore, it is possible to theorize that a tardyon may travel to the future in the retarded wave or that a tachyon may travel to the past in an advanced wave. It is the six-dimensionality of a dynamic space-time that endows the spatiotemporal medium with the attributes of advanced or retarded waves.

1

$$|\Psi^+\rangle = a_x \begin{pmatrix} s_x \\ 0 \\ 0 \end{pmatrix} + a_y \begin{pmatrix} 0 \\ s_y \\ 0 \end{pmatrix} + a_z \begin{pmatrix} 0 \\ 0 \\ s_z \end{pmatrix} \tag{1.1}$$

$$\langle \Psi^- | = a_x^* \left(ct_x, 0, 0 \right) + a_y^* \left(0, ct_y, 0 \right) + a_z^* \left(0, 0, ct_z \right) \tag{1.2}$$

Spatiotemporal Density Matrix or Spatiotemporal Density Operator (ket-bra notation):

A Spatiotemporal Density Operator is a density matrix for the multiplication of a retarded wave with its advanced wave at an arbitrary spatiotemporal point. A density matrix describes the statistical spatiotemporal state of a pure or a mixed quantum mechanical system. The probability for any outcome of any well-defined measurement upon an embedded quantum mechanical system can be calculated from the density matrix of the statistical spatiotemporal state of the system.

$$|\Psi^+\rangle\langle\Psi^-| = \begin{vmatrix} a_x a_x^* & a_x a_y^* & a_x a_z^* \\ a_y a_x^* & a_y a_y^* & a_y a_z^* \\ a_z a_x^* & a_z a_y^* & a_z a_z^* \end{vmatrix} \tag{1.3}$$

§ 2. How would the words of Hermann Minkowski change today with respect to space-time?

The strength of the view of space-time as laid out by Hermann Minkowski which sprung from the soil of experimental physics during his time, would radically evolve into an auspicious concept that preserves the independent reality of the dimensions of space and time as indistinguishable aspects of the same quintessential substance that pervades our universe.

§ 3. What are the quantum mechanical aspects of the probability of the wave function?

The imaginary number "i" represents the complex property, the nature of space-time, or the wave nature of all there is in physical reality. However, it is interesting to know that it also represents the probabilistic aspects of the wave function that all particles, matter, energy, and space-time, with wave properties, follow through the evolution of a system.

The ancient Babylonians took the square root of the radius of a circle three times to calculate an approximate value of π equal to 3. The great mathematician Archimedes of Syracuse calculated the value of π during the third century BC. The eminent mathematician Leonhard Euler introduced the current symbol for π, also known as Archimedes constant, in the eighteenth century AD.

The imaginary number "i" is equal to $\sqrt{-1}$, and its square "i^2" is as probabilistic as "i". The transcendental and irrational number "e" raised to the i-th power is a complex number.

$$e^i = (-1)^{\frac{1}{\pi}} \qquad (3.1)$$

$$\left(e^i\right)^\pi = \left[(-1)^{\frac{1}{\pi}}\right]^\pi \qquad (3.2)$$

$$e^{i\pi} = -1 \qquad (3.3)$$

Let us find the cube roots of Euler's equation when $\theta = \dfrac{\pi}{3}$,

$$\left(e^{i\left(\frac{\pi}{3}\right)}\right)^3 = (-1)^3 \qquad (3.4)$$

$$\left(e^{i\left(\frac{\pi}{3}\right)}\right)^3 + (1)^3 = 0 \qquad (3.5)$$

$$\left(e^{i\left(\frac{\pi}{3}\right)} + 1\right)\left[\left(e^{i\left(\frac{\pi}{3}\right)}\right)^2 - e^{i\left(\frac{\pi}{3}\right)} + 1\right] = 0 \qquad (3.6)$$

Using the quadratic equation, we find three cube roots to Euler's equation,

$$e^{i\left(\frac{\pi}{3}\right)} = -1 \tag{3.7}$$

$$e^{i\left(\frac{\pi}{3}\right)} = 0.5 \pm i0.866025404 = \frac{1}{2} \pm i\frac{\sqrt{3}}{2} \tag{3.8}$$

$$e^{i\left(60^0\right)} = \cos 60^0 \pm i\operatorname{Sin} 60^0 = \frac{1}{2} \pm i\frac{\sqrt{3}}{2} \tag{3.9}$$

Finding the cube roots for $\theta = \dfrac{2\pi}{3}$,

$$e^{i(2\pi)} = 1 \tag{3.10}$$

$$\left(e^{i\left(\frac{2\pi}{3}\right)}\right)^3 = (1)^3 \tag{3.11}$$

The three cube roots are,

$$e^{i\left(\frac{2\pi}{3}\right)} = 1 \tag{3.12}$$

$$e^{i\left(\frac{2\pi}{3}\right)} = -0.5 \pm i0.866025404 = -\frac{1}{2} \pm i\frac{\sqrt{3}}{2} \tag{3.13}$$

$$e^{i\left(120^0\right)} = -\cos 120^0 \pm i\operatorname{Sin} 120^0 = -\frac{1}{2} \pm i\frac{\sqrt{3}}{2} \tag{3.14}$$

These cube roots are the probable locations of a point particle as it translates in its wave function through planes or surfaces at intervals of $\theta = 60^0$, or at intervals of $\theta = 120^0$, in complex space-time. The eminent physicist Christiaan Huygens in the 17th century allegedly said, "I believe we do not know anything for certain, but everything probably." Among other things, Huygens founded the theory of light as an expanding spherical wave. (Huygens, 1690)

The imaginary and transcendental number $e^{i\pi}$ can be the root of an exponential polynomial with imaginary, transcendental, and irrational coefficients, $i^2 - 2i\sqrt{e^{i\pi}} + \left(\sqrt{e^{i\pi}}\right)^2 = 0$, for $i = \sqrt{e^{i\pi}}$.

Thus, $e^{i\pi}$ is also the perfect imaginary and transcendental square, $i^2 = e^{i\pi}$.

Let us denote the following values of "i", "π" and "e",

$$i^2 = -1 = 1\angle 180^0 = e^{i\pi} \qquad (3.15)$$

$$i = \sqrt{e^{i\pi}} = \sqrt{-1} \qquad (3.16)$$

$$\sqrt{i} = \sqrt{\sqrt{e^{i\pi}}} = \sqrt[4]{e^{i\pi}} \qquad (3.17)$$

$$-i^2 = 1\angle 0^0 = 1\angle 180^0 \cdot 1\angle 180^0 = 1\angle 360^0 = -e^{i\pi} \qquad (3.18)$$

$$i^3 = -i = -\sqrt{e^{i\pi}} \qquad (3.19)$$

The transcendental number π is the exponent of the spatiotemporal proportionality of waves.

If a spatial or a temporal variable is raised to the i-th power or multiplied by $i!$ ("i" factorial), the variable would be associated with the properties of a spatiotemporal wave.

$$i^i = \left(e^{\ln(i)}\right)^i = e^{i \cdot \ln(i)} = e^{i \cdot i \cdot \frac{\pi}{2}} = e^{-\frac{\pi}{2}} = \frac{1}{e^{\frac{\pi}{2}}} \approx \frac{1}{5} \qquad (3.20)$$

$$\ln i^i = -\frac{\pi}{2} \qquad (3.21)$$

Denoting "π" in terms of "i" and "e", we obtain

$$\pi = -2\ln i^i \qquad (3.22)$$

$$\sqrt[2]{\pi} = \sqrt[2]{-2\ln i^i} \qquad (3.23)$$

An expression for the three dimensional proportionality of π is given by

$$\sqrt[3]{\pi} = \sqrt[3]{-2\ln i^i} = (i)\sqrt[3]{2\ln i^i} = (i)\sqrt[3]{\ln i^{2i}} \qquad (3.24)$$

Euler's equation in terms of "i" and "e" only, may be expressed as

$$e^{-(2i)\ln(i^i)} = -1 \qquad (3.25)$$

The "i" factorial yields the complex number of a spatiotemporal wave. A damped spatiotemporal wave whose amplitude of oscillation decreases with time, eventually going to zero, an exponentially decaying wave.

$$i! = \int_0^\infty t^i e^{-t} dt = \int_0^\infty \left(e^{\ln t}\right)^i e^{-t} dt = \int_0^\infty \left(e^{i\ln t}\right) e^{-t} dt \qquad (3.26)$$

$$i! = \int_0^\infty e^{-t} \mathrm{Cos}(\ln t) dt + \int_0^\infty e^{-t} \mathrm{Sin}(\ln t) dt \qquad (3.27)$$

$$i! \approx \frac{1}{2} - i\frac{2}{13} \qquad (3.28)$$

It is interesting to note that the double factorial of "i" still renders the wave property.

$$i!! = 2^{\frac{i}{2}}\left(\frac{i}{2}\right)! \qquad (3.29)$$

The transcendental numbers, "e", "π", and "i", are correlated in the expansion or contraction of the spatiotemporal wave function, as

6

spatial and temporal distances change according to the natural base "e", the phase angle of the spatiotemporal wave gradually incrementing according to the proportionality ratio "π", and the imaginary number "i" manifesting the property of a wave as the kernel of growth changes over time. The proportionality aspect and the wave property aspect have an exponential effect on the natural base of growth.

Let us represent the probability principle for the wave function of the particle-wave duality,

$$\sqrt{i} = \sqrt{(\lambda + i\theta)^2} \tag{3.30}$$

$$i = \lambda^2 + (2\lambda\theta)i - \theta^2 \tag{3.31}$$

Since both $\theta = 0$ and $\lambda = 0$, we have

$$2\lambda\theta = 1 \tag{3.32}$$

$$\theta = \frac{1}{2\lambda} \tag{3.33}$$

Because $\lambda^2 - \theta^2 = 0$, we can find the value of λ,

$$\lambda^2 - \left(\frac{1}{2\lambda}\right)^2 = 0 \tag{3.34}$$

$$\lambda = \pm\frac{1}{\sqrt{2}} \tag{3.35}$$

Substituting for λ to find θ,

$$\theta = \frac{1}{2\lambda} = \pm\frac{\sqrt{2}}{2} = \pm\frac{1}{\sqrt{2}} \tag{3.36}$$

$$\sqrt{i} = \pm\lambda \pm \theta i \tag{3.37}$$

7

$$\sqrt{i} = \pm \frac{1}{\sqrt{2}} \pm \frac{1}{\sqrt{2}} i \qquad (3.38)$$

$$i = \left(\pm \frac{1}{\sqrt{2}} \pm \frac{1}{\sqrt{2}} i \right)^2 \qquad (3.39)$$

Using Euler's equation to verify the previous result,

$$-1 = i^2 = e^{i\pi} \qquad (3.40)$$

$$i = \pm \sqrt{e^{i\pi}} \qquad (3.41)$$

$$\sqrt{i} = \pm \sqrt{\sqrt{e^{i\pi}}} = \pm e^{i\frac{\pi}{4}} = \pm e^{i45^0} \qquad (3.42)$$

$$\sqrt{i} = \pm e^{i45^0} = \pm \cos 45^0 \pm i \operatorname{Sin} 45^0 = \pm \frac{1}{\sqrt{2}} \pm \frac{1}{\sqrt{2}} i \qquad (3.43)$$

It is interesting to observe that $\pi/4$ or $\left[(\tfrac{1}{2})! \right]^2$ is the phase angle of the spatiotemporal trajectory of a particle that has equal probability to be either a particle or a particle-wave manifestation in physical reality. The phase of a repetitive waveform specifies the location of a point within a wave cycle. The kernel of growth may also represent the natural wave operator, $\pm e^{i\theta}$, for a wave traveling in either direction of a dimension, and the kernel of the probability of Quantum Mechanics. The probability property is conserved even if space-time is expanding or contracting, in which case, the exponent of "e" would be a complex number, $\pm e^{\pm \ln \psi \pm i\theta}$, where Ψ is the amplitude of the wave, and θ is the phase angle.

The product rule for the nth root of a negative number or an imaginary number states that the non-linear nth root operator "$\sqrt[n]{}$" for any natural number "n" has both a positive sign and a negative sign in front of the nth root, "$\pm \sqrt[n]{}$". Consequently, whenever we take the nth root of a negative number or an imaginary number, the result could be either positive or negative. Therefore, when two or more nth roots of a negative number, or of an imaginary number, are multiplied, the result could be either positive or negative, depending on the ultimate sign of the product of all operands under the nth root operator to avoid a fallacy. Thus, the chosen branch of the result has to match the ultimate sign on the other side of the equation or equality that is equal to the nth root operation.

Applying the non-linear nth root operator to imaginary numbers,

If $n = 1$, we have

$$\sqrt[i]{i} = i^{-i} = \frac{1}{i^i} = \frac{1}{e^{-\frac{\pi}{2}}} = e^{\frac{\pi}{2}} \tag{3.44}$$

$$\sqrt[i]{i^2} = \left(i^2\right)^{-i} = \frac{1}{i^{2i}} = \frac{1}{i^i} \cdot \frac{1}{i^i} = \frac{1}{e^{-\frac{\pi}{2}}} \cdot \frac{1}{e^{-\frac{\pi}{2}}} = \frac{1}{e^{-\pi}} = e^{\pi} \tag{3.45}$$

If $n = 2$, we obtain

$$\sqrt[i^2]{i} = (i)^{\frac{1}{i^2}} = i^{-1} = \frac{1}{i} = \frac{1}{i} \cdot \frac{i}{i} = \frac{i}{i^2} = -i \tag{3.46}$$

$$\sqrt[i^2]{i^2} = \left(i^2\right)^{\frac{1}{i^2}} = (-1)^{-1} = \frac{1}{(-1)} = -1 = e^{i\pi} \tag{3.47}$$

It is interesting to observe that the results may be real or imaginary numbers, including the kernel of growth.

Let us introduce the non-linear imaginary nth root operator as "$\pm \sqrt[i^n]{\ }$" for any natural number "n" as the exponent of the imaginary base "i" of the logarithm. The operand could be an imaginary and/or a transcendental number. The result could be either a positive or a negative imaginary and/or transcendental number. The non-linear imaginary nth root operator may be applied to a temporal distance, to the property of a wave, or to the probability of a quantum mechanical object.

For isotropic and homogeneous space-time, the three-dimensional temporal probability operator "i" is given by

$$i^2 = i_x^2 + i_y^2 + i_z^2 \tag{3.48}$$

$$i_x = i_y = i_z = \frac{i}{\sqrt[2]{3}} \tag{3.49}$$

Describing the wave function in terms of "i" and the kernel of natural growth e^{-r}, we obtain

$$\left[\Psi(r)\right]^2 = \frac{i^2}{\ln e^{-r}} = \frac{\left(\pm\dfrac{1}{\sqrt{2}} \pm \dfrac{1}{\sqrt{2}}i\right)^4}{-r} \tag{3.50}$$

$$\frac{\left[\Psi(r)\right]^2}{i} = \frac{i}{\ln e^{-r}} \tag{3.51}$$

$$-i\left[\Psi(r)\right]^2 = \frac{i}{\ln e^{-r}} = \frac{\left(\pm\dfrac{1}{\sqrt{2}} \pm \dfrac{1}{\sqrt{2}}i\right)^2}{-r} \tag{3.52}$$

$$\left[\Psi(r)\right] = \frac{\pm\dfrac{1}{\sqrt{2}} \pm \dfrac{1}{\sqrt{2}}i}{\sqrt{ir}} = \frac{\sqrt{i}}{\sqrt{i}\sqrt{r}} = \frac{1}{\sqrt{r}} \tag{3.53}$$

$$\Psi(r) = \frac{1}{\sqrt{r}} \tag{3.54}$$

$$\left|\Psi(r)\right|^2 = \frac{1}{r} \tag{3.55}$$

$$\left|\Psi(r)\right|^4 = \frac{1}{r^2} \tag{3.56}$$

10

How is the wave function related to Euler's equation?

$$\lim_{r \to \infty} \left(1 + \frac{1}{\sqrt{r}}\right)^{\sqrt{r}} = e \qquad (3.57)$$

$$\lim_{\frac{1}{|\Psi(r)|^2} \to \infty} \left[1 + \Psi(r)\right]^{\frac{1}{\Psi(r)}} = e \qquad (3.58)$$

$$\lim_{\frac{1}{|\Psi(r)|^2} \to \infty} \left[1 + \Psi(r)\right]^{\frac{1}{\Psi(r)}} = \lim_{\frac{1}{|\Psi(r)|^2} \to \infty} \left[1 + |\Psi(r)|^2\right]^{\frac{1}{|\Psi(r)|^2}} \qquad (3.59)$$

$$= \lim_{\frac{1}{|\Psi(r)|^2} \to \infty} \left[1 + |\Psi(r)|^4\right]^{\frac{1}{|\Psi(r)|^4}} = e$$

$$\lim_{\frac{1}{|\Psi(r)|^2} \to \infty} \left(\left[1 + |\Psi(r)|^4\right]^{\frac{1}{|\Psi(r)|^4}}\right)^{\left(\frac{i}{|\Psi(r)|^4}\right)^{\theta}} = e^{i\theta} = \operatorname{Cos}\theta + i\operatorname{Sin}\theta \qquad (3.60)$$

Therefore, as the wave function, or variables of wave property, are multiplied by "i", the attribute of probability is mathematically associated with the wave function or wave property as it is in physical phenomena. From the last equation of spatial curvature, it is interesting to observe how the wave property emerges from the limit of the wavefunction expression, starting from the initial condition of what is there "e^0" plus the wavefunction, as the spatial length expands to infinity, as time endows space, then space endows more time. The wavefunction is the natural base "e" of growth, which is itself a wave for all natural systems. The angle "θ" is the phase angle of the exponential natural growth.

In the case of particle-wave duality, the particle manifestation has a fifty percent probability, and the particle-wave manifestation has an equal fifty percent property of occurring in physical reality, the outcomes are equally divided, depending on the physical variable of a possible set that defines a physical system or sets the conditions of its operation.

The probability principle asserts space-time and all physical manifestation of energy as emergent properties of complex physical reality. The spatiotemporal emergence endows the wave function with its probabilistic property. Time emerges to create more space, which in turn creates more time. Thus, probability, just like space-time, is complex, it consists of real probability (spatial) and imaginary probability (temporal). Also, $t_p \equiv il_p$, $i \equiv t_p/l_p$, *therefore*, $i \equiv 1/c_p$. The plus or minus of the probability of "i" are the spatial or the temporal directions or senses of a spatial dimension and its "complex conjugate" temporal dimension. Space or time may expand or contract as space-time emerges.

§ 4. A spatiotemporal point source.

If every spatiotemporal point source is considered as a traversable spatiotemporal bridge at quantum scale from the past to the present under present mathematical models that allow this hypothesis, during the expansion of the universe just after the big bang, these spatiotemporal bridges would have been stretched out and may have been stabilized by spatiotemporal strings. Such traversable spatiotemporal bridges would allow particles to travel back to the very beginning of our universe. One such purely geometric spatiotemporal bridge is the Morris-Thorne spatiotemporal bridge with no horizons, no singularities, perfectly stable in time, not closing in on itself while a particle or object traverses it, and not generating a gravitational field around it. A spaceship could enter and exit on the same side through geodesic trajectories. It has a lensing effect on light rays passing near it, and it could serve as a telescope to view the orientation and image on other side of the bridge. This attribute could be used as a very powerful telescope of the universe for the far field. The spatiotemporal tension of the spatiotemporal bridge has to be counteracted by a force from a negative source of energy. By the topological censorship theorem, a negative source of energy is not allowed under the present perspective of the General Theory of Relativity.

$$ds^2 = c^2 dt^2 - d\rho^2 - \left(\rho^2 + R^2\right)d\Omega^2 \qquad (4.1)$$

The flaring-out condition entails the violation of the Null Energy Condition (NEC: $\rho + p_r \geq 0$) at the throat: (negative energy densities not essential):

$$\rho + p_r < 0 \tag{4.2}$$

In fact, it violates all of the pointwise energy conditions, the averaged energy conditions, quantum inequalities, semi-classical energy conditions, etc.

Note that the null energy condition arises when one refers back to the Raychaudhuri equation: the positivity condition of the expansion term appears:

$$\rho = T_{\mu\nu}kk^\nu \geq 0 \tag{4.3}$$

In General Relativity, through the EFE the positivity condition reflects the Null Energy Condition (NEC):

$$T_{\mu\nu}k^\mu k^\nu \leq 0 \tag{4.4}$$

(not allowed in General Relativity presently)

$$d\sigma^2 = c^2 dt^2 - ds^2 \tag{4.5}$$

General Equation of a Spatiotemporal Bridge:

A mathematical equation motivated by the symmetry and geometry of cones for the maximally extended Schwarzschild solution. The Kruskal–Szekeres coordinates of General Relativity are well-behaved everywhere outside the physical singularity and they cover the entire spatiotemporal manifold of the maximally extended Schwarzschild solution.

$$u^2 - v^2 = \left(\frac{r}{R_S} - 1 \right) e^{\frac{r}{R_S}} \tag{4.6}$$

$$u = \pm \sqrt{v^2 + \left(\frac{r}{R_S} - 1\right) e^{\frac{r}{R_S}}} \qquad (4.7)$$

$$\tanh\left(\frac{t}{4GM}\right) = \begin{cases} v/u, & r > 2GM \\ u/v, & r < 2GM \end{cases} \qquad (4.8)$$

The entrance and the exit cones of the Schwarzschild spatiotemporal bridge may be expressed as the positive or the negative square root of the equation.

Where R_S is the Schwarzschild radius, and r, u and v are Kruskal coordinates over space-time on a Kruskal diagram of the Schwarzschild geometry.

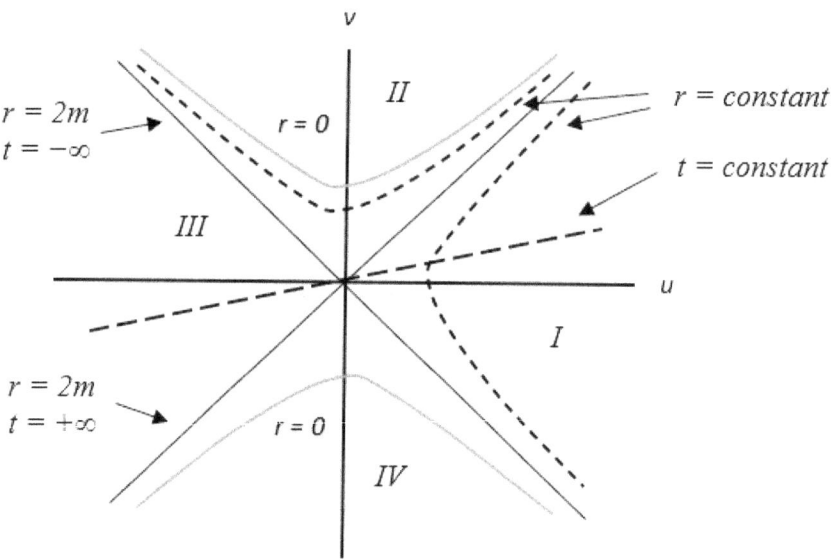

Figure 1. A Kruskal Diagram of the Schwarzschild Geometry

§ 5. The spatiotemporal wave function at a point source.

Are the attributes of the Special or the General Theory of Relativity emergent from the quantum scale? How do the attributes

14

of the Special or the General Theory of Relativity emerge from a Quantum Theory like "A Dynamic Theory of Space-Time: A Matter of Waves"?

Let us describe the spatiotemporal wave function as it emerges at a quantum level from a point source. If the wave function is not obstructed, it would be spherical in isotropic and homogeneous space-time, otherwise, it could be an elliptical wave function due to, but not limited to, the presence of mass, matter, energy, or translation through space. As the wave is obstructed, spatiotemporal pressure increases at the local region of obstruction where the spatial length contracts and the temporal distance dilates according to the Special Theory of Relativity.

Let us imagine an ellipsoidal plane wave that expands through a spatial major axis "s_x" and a perpendicular temporal minor axis "t_x", which may be visualized as an ellipse around the center point, or origin, of an oblate spheroid. The ellipsoidal plane wave may represent a wave expanding through any two rectangular coordinate axes, also from the center point of an expanding oblate spheroid in six-dimensional space-time.

In an ellipse, the ratio between the linear eccentricity and the semi-major axis is a constant. When the eccentricity of the ellipse changes so does the ratio. As an ellipsoidal wave is contracted further its semi-major axis extends and its semi-minor axis contracts while the linear eccentricity extends.

Let us consider the ellipsoidal wave function through the x-axis and the y-axis to represent the two-dimensional aspects of expanding space-time. As the wave expands or contracts, the spatial major axis and the temporal minor axis are reciprocal in their actions, but proportional to the circumference of the ellipse in isotropic and homogeneous space-time with negligible curvature. Using the semi-major axis as coordinate time $d\pi$, the linear eccentricity distance between the origin and the right focus point "F_2" as proper time, and the semi-minor axis as spatial distance $d\Sigma$, of the wave as it expands or contracts, the reciprocal relation can be expressed as the Lorentz factor "gamma" of special relativity, for the inherent geometrical mechanism of the ellipsoidal plane wave. *The reciprocal relation of the spatial and temporal dimensions of the complex wave function is the essence of Special Relativity.*

By the Pythagorean Theorem,

$$d\tau^2 = d\pi^2 - \frac{d\Sigma^2}{c^2} \tag{5.1}$$

Dividing by coordinate time $d\pi$, we get the velocity of proper time, and the gamma factor, which is coordinate time over proper time, $d\pi/d\tau$.

$$v_\tau^2 = \frac{d\tau^2}{d\pi^2} = \frac{d\pi^2}{d\pi^2} - \frac{d\Sigma^2}{d\pi^2 c^2} = 1 - \frac{d\Sigma^2/d\pi^2}{c^2} = 1 - \frac{v_s^2}{c^2} \tag{5.2}$$

$$v_\tau = \sqrt{1 - \frac{v_s^2}{c^2}} \tag{5.3}$$

$$\gamma = \frac{1}{v_\tau} = \frac{d\pi}{d\tau} = \frac{1}{\sqrt{1 - \frac{v_s^2}{c^2}}} \tag{5.4}$$

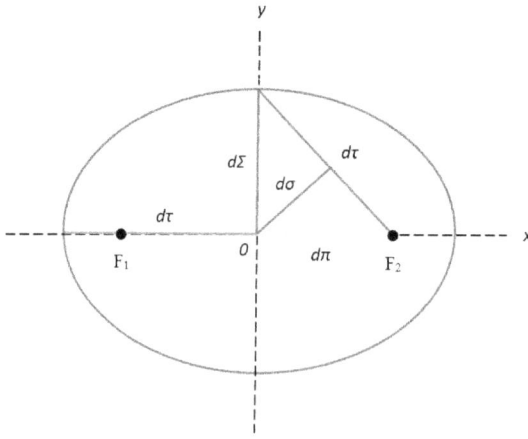

Figure 2. Diagram of a Spatiotemporal Ellipsoidal Wave

By the Inverse Pythagoras Theorem,

$$\frac{1}{d\pi^2} + \frac{c^2}{d\Sigma^2} = \frac{c^2}{d\sigma^2} \tag{5.5}$$

$$\frac{d\sigma^2}{d\pi^2} + c^2 \frac{d\sigma^2}{d\Sigma^2} = c^2 \tag{5.6}$$

$$v_\tau^2 = c^2 - c^2 \frac{d\sigma^2}{d\Sigma^2} \tag{5.7}$$

$$\frac{v_\tau^2}{c^2} = 1 - \frac{d\sigma^2}{d\Sigma^2} \tag{5.8}$$

$$\frac{d\sigma^2}{d\Sigma^2} = 1 - \frac{v_\tau^2}{c^2} \tag{5.9}$$

Where v_σ is the velocity of proper space (geodesic space); geodesic space is a length space,

$$v_\sigma = \sqrt{1 - \frac{v_\tau^2}{c^2}} \tag{5.10}$$

The Larmor factor is given by

$$\Sigma = \frac{1}{v_\sigma} = \frac{1}{\sqrt{1 - \frac{v_\tau^2}{c^2}}} \tag{5.11}$$

The following equation represents the principle of spatiotemporal equivalence for the topology of the spatiotemporal ellipsoidal wave, or the Lemaître principle of relativistic wave topology.

$$\frac{d\Sigma \cdot d\pi}{d\sigma} = d\tau \tag{5.12}$$

$$d\Sigma \cdot d\pi = d\sigma \cdot d\tau \qquad (5.13)$$

$$a = \frac{d\sigma \cdot d\tau}{d\Sigma \cdot d\pi} \qquad (5.14)$$

The product of coordinate space and coordinate time is equal to the product of proper space (geodesic space) and proper time. The coordinate spatial distance and the coordinate temporal distance are co-moving dimensions, that is, they expand or contract with the expansion or contraction of our universe. The proper spatial distance (proper space) and the proper temporal distance (proper time) are the spatiotemporal distances that we would normally measure in our reality. The scale factor of the expansion of the universe "a" is the ratio of proper space-time over co-moving space-time.

Distant galaxies remain at approximately the same spatiotemporal radial co-moving distance (peculiar distance) from each other as space-time expands (as doppler redshift approaches "c"), but they move farther apart (recessive distance can expand faster than light as a change of scale over time) if they are separated from each other at the same radial distance from the point of observation (cosmological redshift may be faster than light). The Hubble (radial) velocity is approximately 43.5 *miles/sec/Mpc* (70 *Km/s/Mpc*). The total distance of expansion between two galaxies would be the sum of the recessive distance and the peculiar distance.

$$\frac{1}{\dfrac{d\sigma}{d\Sigma}} = \frac{d\tau}{d\pi} \qquad (5.15)$$

$$\frac{1}{v_\sigma} = v_\tau \qquad (5.16)$$

$$\frac{1}{\dfrac{\partial v_\sigma}{\partial t}} = \frac{\partial v_\tau}{\partial t} \qquad (5.17)$$

$$\frac{1}{a_\sigma} = a_\tau \qquad (5.18)$$

The velocity of proper space (geodesic space) is reciprocal to the velocity of proper time. Therefore, the acceleration of proper space is reciprocal to the acceleration of proper time.

The wavefunction of Quantum Mechanics may be defined as

$$\frac{1}{d\Sigma} = \frac{d\pi}{d\sigma \cdot d\tau} = \frac{v_\tau}{d\sigma} \tag{5.19}$$

$$\frac{1}{\sqrt{d\Sigma}} = \sqrt{\frac{v_\tau}{d\sigma}} = \sqrt{\frac{\sqrt{1-\frac{v_r^2}{c^2}}}{d\sigma}} = \frac{\sqrt[4]{1-\frac{v_r^2}{c^2}}}{\sqrt{d\sigma}} \tag{5.20}$$

$$\Psi(\Sigma) = \frac{1}{\sqrt{d\Sigma}} = \sqrt{\frac{v_\tau}{d\sigma}} = \frac{\sqrt[4]{1-\frac{v_r^2}{c^2}}}{\sqrt{d\sigma}} \tag{5.21}$$

$$\left|\Psi(\Sigma)\right|^2 = \frac{1}{d\Sigma} = \frac{|v_\tau|}{d\sigma} = \frac{\sqrt{1-\frac{v_r^2}{c^2}}}{d\sigma} \tag{5.22}$$

The spatial wave function is relativistic as a function of proper time and proper space. The relativistic wavefunctions can be expressed in integral form as the spatial and temporal transforms of the Special or the General Relativity.

$$\Psi(r) = \frac{1}{\sqrt{2\pi}} \int_{-\infty}^{\infty} \Psi(t) \cdot e^{-st} \cdot dt \tag{5.23}$$

$$\Psi(t) = \frac{1}{\sqrt{2\pi}} \int_{-\infty}^{\infty} \Psi(r) \cdot e^{-sr} \cdot dr \tag{5.24}$$

where the complex frequency is $s = \sigma + i\omega = r\angle\theta$, with real numbers σ and ω.

Expressing the relativistic spatiotemporal wave function, we obtain

19

$$\Psi(r,t) = \frac{1}{2\pi} \int_{-\infty}^{\infty} \int_{-\infty}^{\infty} \Psi(r) e^{-s(r+t)} dr \cdot dt \qquad (5.25)$$

It is interesting to note that the magnitude and the phase angle of the complex frequency "s", in the exponent of the kernel of growth "e" multiplying the spatial wavefunction "$\Psi(r)$", plays a crucial role to determine the expansion or contraction of the spatiotemporal wavefunction since spatial length "r" and temporal distance "t" are reciprocal.

§ 6. Is it possible to express spatiotemporal growth using Euler's formula?

It is possible to express a relativistic spatiotemporal growth of $i^2 e^{\pi+i\pi}$ in the space-time-energy medium, using Euler's formula which is common in engineering, mathematics, and physics.

$$e^{i\pi} = \mathrm{Cos}\ \pi + i\,\mathrm{Sin}\ \pi = -1 \qquad (5.26)$$

$$e^{i\pi} \mp ie^{i\pi} = e^{i\pi}\left(1 \mp i\right) = -1 \pm i \qquad (5.27)$$

$$-e^{i\pi} = 1 \mp i \qquad (5.28)$$

$$1 \mp i \equiv 1 \pm i \qquad (5.29)$$

$$e^{\pi} \pm ie^{\pi} = e^{\pi}\left(1 \pm i\right) = e^{\pi} \cdot \left(-e^{i\pi}\right) = -e^{\pi+i\pi} = i^2 e^{\pi+i\pi} \qquad (5.30)$$

The Lorentz factor "gamma" used in the Special Theory of Relativity is given by

$$\gamma = \frac{1}{\sqrt[2]{1 - \dfrac{v^2}{c^2}}} = \frac{dt}{d\tau} \qquad (5.31)$$

Where "t" is coordinate time, "τ" is proper time, "v" is the velocity of an object, and "c" is the speed of light.

$$\gamma = \cosh\left(i^2 e^{\pi+i\pi}\right) = \frac{1}{\sqrt[2]{1-\tanh^2\left(i^2 e^{\pi+i\pi}\right)}} \qquad (5.32)$$

$$\tanh^2\left(i^2 e^{\pi+i\pi}\right) = -\tanh^2\left(e^{\pi+i\pi}\right) = \frac{v^2}{c^2} \qquad (5.33)$$

$$\frac{v}{c} = \tanh\left(i^2 e^{\pi+i\pi}\right) = -\left(\frac{e^{(\pi+i\pi)}-e^{-(\pi+i\pi)}}{e^{(\pi+i\pi)}+e^{-(\pi+i\pi)}}\right) \qquad (5.34)$$

$$i^2 e^{(\pi+i\pi)} = -\sqrt[2]{\frac{c+v}{c-v}} \qquad (5.35)$$

$$i^2 e^{-(\pi+i\pi)} = -\sqrt[2]{\frac{c-v}{c+v}} \qquad (5.36)$$

$$\frac{dt}{d\tau} = \frac{1}{\sqrt[2]{1-\frac{v^2}{c^2}}} = \frac{1}{\sqrt[2]{1-\left(\frac{e^{(\pi+i\pi)}-e^{-(\pi+i\pi)}}{e^{(\pi+i\pi)}+e^{-(\pi+i\pi)}}\right)^2}} \qquad (5.37)$$

$$= \frac{1}{\sqrt[2]{\left(\frac{e^{(\pi+i\pi)}+e^{-(\pi+i\pi)}}{e^{(\pi+i\pi)}+e^{-(\pi+i\pi)}}\right)^2 - \left(\frac{e^{(\pi+i\pi)}-e^{-(\pi+i\pi)}}{e^{(\pi+i\pi)}+e^{-(\pi+i\pi)}}\right)^2}}$$

As stated in my book "A Dynamic Theory of Space-Time: A Matter of Waves", it is interesting to note that the first term that multiplies dt^2 indicates the condition during spatiotemporal expansion that equals "1", when the advanced wave is offset by the retarded wave. The second term indicates the condition during spatiotemporal expansion that equals a fraction, when there is a remainder wave after interference of the advanced and retarded waves. The difference between the two terms is also a fraction that when multiplied by the squared coordinate time, dt^2, is equal to the squared proper time, $d\tau^2$. Therefore, the above equation for the interference of

21

spatiotemporal waves demonstrates a possible explanation for spatiotemporal General Relativity when an object of mass moves in a spatiotemporal direction at a speed less than the speed of light.

The above mathematical equations confirm that *A Quantum Principle of the Special Relativity* is inherent in A Dynamic Theory of Space-Time as a Quantum Theory of Gravitation.

Chapter 2

Space-Time and Mass

§ 1. The expansion of space-time

What is the physical reality of the expanding vacuum? What exactly in space-time is expanding? Why does the expansion phenomenon occur? What physical processes are causing the expansion and acceleration of space-time?

What is the scalar curvature in six-dimensional space-time of a star going supernova for the mass and energy density of this phenomenon of nature?

The scalar curvature of a star going "spherical supernova" may be described by the amount of change of its spherical volume in curved space-time from its spherical volume in a Minkowski space.

(Pressure · Speed of Light) ≡ Einsteinian Energy Density

$$G_{\mu v} = \frac{8\pi r}{hg}\left(E_{\mu v} - \tilde{\Lambda}_{\mu v}\right) \qquad (1.1)$$

Where "h" is the Planck constant, "g" is the gravitational acceleration, "r" is the radius of the spherical supernova, $E_{\mu v}$ is the Einsteinian six-dimensional stress-energy-momentum tensor, and $\tilde{\Lambda}_{\mu v}$ is the cosmological Einsteinian six-dimensional stress-energy-momentum tensor.

$$R_{\mu v} - \frac{1}{(n-1)}g_{\mu v}R = \frac{8\pi r}{hg}\left(E_{\mu v} - \tilde{\Lambda}_{\mu v}\right) \qquad (1.2)$$

Using $n = 6$ for six spatiotemporal dimensions, "ω" is the angular frequency, P_E for Einsteinian energy density, and P_E for Einsteinian pressure, we obtain

$$R_{\mu\nu} - \left(\frac{1}{5}\right)g_{\mu\nu}R = -\frac{8\pi}{h\omega^2}\left(-3\rho_E + 3p_E\right) \qquad (1.3)$$

From previous research, it was found that the metric tensor may be obtained from the cosmological matter and energy, and the cosmological gravitational waves from the gravitational field present through a vacuum region of space-time whose limit is the Newtonian potential. (Nieves, 2020)

Since $g^{\mu\nu}g_{\mu\nu} = 6$ in a nearly-flat six-dimensional Riemannian manifold, we have

$$\left[\left(\frac{5}{5}\right)R - \left(\frac{6}{5}\right)R\right] = -\frac{8\pi r}{hg}\left(-3\rho_E + 3p_F\right) \qquad (1.4)$$

$$-\left(\frac{1}{5}\right)R = -\frac{8\pi r}{hg}\left(-3\rho_E + 3p_E\right) \qquad (1.5)$$

$$-R = -\frac{40\pi r}{hg}\left(-3\rho_E + 3p_E\right) \qquad (1.6)$$

$$6\left(\frac{\ddot{a}}{ac^2} + \frac{\dot{a}^2}{a^2c^2} + \frac{k}{a^2}\right) = \frac{40\pi r}{hg}\left(-3\rho_E + 3p_E\right) \qquad (1.7)$$

$$\left(\frac{\ddot{a}}{ac^2} + \frac{\dot{a}^2}{a^2c^2} + \frac{k}{a^2}\right) = \frac{20\pi r}{hg}\left(-\rho_E + p_E\right) \qquad (1.8)$$

$$\left(\frac{\ddot{a}}{ac^2} + \frac{\dot{a}^2}{a^2c^2} + \frac{k}{a^2}\right) = \frac{10r}{\hbar g}\left(-\rho_E + p_E\right) \qquad (1.9)$$

Reformulating the six-dimensional field equation for a spherical supernova, we have

$$R_{\mu\nu} - \frac{1}{(n-1)}g_{\mu\nu}R = \frac{4r}{\hbar g}\left(E_{\mu\nu} - \tilde{\Lambda}_{\mu\nu}\right) \qquad (1.10)$$

If the body of mass *"m"* has a charge *"q"*, we find

$$\frac{r}{hg} = \frac{1}{mass \cdot c} \cdot \frac{1}{g} = \frac{1}{mass \cdot c} \cdot \left(\frac{s^2}{m} \cdot \frac{m^2}{m^2} \right)$$ (1.11)

$$= \frac{1}{mass \cdot c} \cdot \left(\frac{q^2}{Vol.} \right) = \frac{q}{mass \cdot c} \cdot \frac{q}{Vol.}$$

$$\frac{r}{hg} = \frac{q}{mass \cdot c} \cdot Q$$ (1.12)

Where *"Q"* is the proper charge density. Hence, we may express the six-dimensional electrogravitic field equation for a spherical supernova as

$$R_{\mu\nu} - \frac{1}{(n-1)} g_{\mu\nu} R = \frac{8\pi q}{mc} \left(Q_{\mu\nu} \right) \left(E_{\mu\nu} - \tilde{\Lambda}_{\mu\nu} \right)$$ (1.13)

$$R_{\mu\nu} - \frac{1}{(n-1)} g_{\mu\nu} R = \frac{8\pi\omega}{\vec{\Phi}_{\mu\nu}} \left(Q_{\mu\nu} \right) \left(E_{\mu\nu} - \tilde{\Lambda}_{\mu\nu} \right)$$ (1.14)

From previous research, a charge *"q"* (in Coulombs) is defined as the product of a spatial length and a temporal distance, $Q_{\mu\nu}$ is the proper charge density tensor, and $\vec{\Phi}_{\mu\nu}$ is the resultant electric field tensor. (Nieves, 2020)

It is a good time now to ask the rhetorical question, if the proper charge density tensor represents a significant DC charge and the resultant electric field tensor represents a rectified and alternating electric field, so that the outward force of the field per coulomb maintains a constant value of the combined AC/DC electric field about the geometry of the body of mass, would varying the charge density cause a variation of the spatiotemporal curvature about the body of mass?

$$\vec{\Phi}_{\mu\nu} = \vec{\Phi}_{DC} \pm \vec{\Phi}_{AC}$$ (1.15)

§ 2. A discrepancy in the Einstein's Field Equations.

From a filmed interview of Paul A. M. Dirac by Physicist Friedrich Hund circa 1982 in Göttingen at the Institut für den Wissenschaftlichen:

Dirac said that the time and distance calculated by Einstein's equation is not the same as time and distance provided by an atomic clock. Hence, there is a discrepancy between a very precise atomic clock and Einstein's equation. Dirac compared a weakening gravitational force in relation to an electromagnetic force. Dirac said during an interview:

"During recent work I have been very much concerned with Einstein's General Relativity and I believe that the times and distances which are to be used in Einstein's General Relativity are not the same as the times and distances which would be provided by atomic clocks. There are good theoretical reasons for believing that is so and for believing that gravitational forces are getting weaker compared to electric forces as the world gets older. There is some observational evidence for that. Observations of the moon that have been made accurately for centuries with respect to the time provided by the Einstein Theory and which have been made since nineteen hundred and fifty five with atomic clocks and there is some evidence of a difference between the two times. The evidence is not as complete as one would like to have, people are still working on the subject in particular with the Viking lander which was put onto Mars in 1976. One is able to send radar waves to Mars and get back the reflected waves and one can measure in atomic time how long it takes these waves to go to Mars and to come back. The results that one gets are unfortunately very complicated because there are many disturbances there. There are disturbances caused even by meteors. There are many more meteors passing close to Mars that there are passing close by the Earth, and these disturbances all have to be taken into account. Well people are still working on this subject, and I hope they will get a definite answer pretty soon about the question of whether there are these two times, the Einstein time and the atomic time, with a difference between them."

Is this due to the one dimensionality of time and the three

dimensionality of space in Einstein's equations? Has the question of the interpretation of Einstein's General Relativity equations been settled? What spatial ruler would be used with the atomic clock for measurement? How accurate can the atomic spatial measurement be?

The discrepancies may be due to scale. Curvature may change depending on the scale of a measurement of distance or a measurement of time. At the atomic scale near a large body of mass or at a great distance above the surface of a celestial body, time may run faster than at the surface of the massive or celestial body. The arcsecond or radial distance of the atomic clock may have a compressed length, and time may be less dilated. Curvature, the length of spatial distance, the dilation of a dimension of time, are not conformal according to scale. Hence, the measurement of a spatial distance, and the passing of time or the temporal dilation of a temporal dimension, of an astronomical measurement versus an atomic measurement is not conformal according to scale and may lead to a discrepancy in measurement and in a disagreement with the result from the EFEs. The EFEs are to be applied at a local scale for space and time, they are not scale invariant. An additional curvature tensor differential may have to be added in the EFEs to compensate for scale, radial length expansion or compression, and temporal dilation, or for local differences in spatiotemporal pressure as well as torsion.

§ 3. *Would varying the spatiotemporal curvature cause a variation in the gravitational acceleration about the body of mass? The greatest truths of nature do not hide from faultless experiments!*

Even energy may be relativistic; relativistic Kinetic energy is given by

$$K.E. = mc^2 \left(\frac{1}{\sqrt{1 - \dfrac{v_r}{c^2}}} - 1 \right) \qquad (3.1)$$

Can the equation for Gauss's law be derived from the six-dimensional electrogravitic field equation?

27

$$R_{\mu\nu} - \frac{1}{(5)} g_{\mu\nu} R = \frac{8\pi q}{mc}(Q_{\mu\nu})(E_{\mu\nu} - \tilde{\Lambda}_{\mu\nu}) \qquad (3.2)$$

$$-\frac{1}{5} R = \frac{8\pi q}{mc}(Q_{\mu\nu})(E_{\mu\nu} - \tilde{\Lambda}_{\mu\nu}) \qquad (3.3)$$

$$-\frac{1}{5q}\frac{\partial^2(mass \cdot c^3)}{\partial r^2} = 8\pi c\left(\frac{qc^3}{Vol.}\right)\left(\frac{Einsteinian\ Energy}{Vol.}\right) \qquad (3.4)$$

Dividing both sides of the equation by speed of light "c", with $n = 3$ to convert energy from 6 to 4 spatiotemporal dimensions, we obtain

$$-\frac{1}{q}\frac{\partial^2(mass \cdot c^2)}{\partial r^2} = 8\pi\left(\frac{qc^2}{Vol.}\right)\left(\frac{Energy}{Vol.}\right) \qquad (3.5)$$

$$\frac{1}{q}\frac{\partial^2\left(\frac{1}{2}mass \cdot c^2\right)}{\partial r^2} = 4\pi q\left(-\frac{1}{m^3}\right)\left(\frac{J}{m^3}\right)\left(\frac{r^2}{t^2}\right) \qquad (3.6)$$

Where "F" is a four-dimensional force of divergence and "a" is an acceleration, c^2/r.

$$\frac{1}{q}\int\frac{\partial\left(ma\frac{dr}{dt}\right)}{\partial r}dt = 4\pi q\left(-\frac{1}{m^2}\right)\left(\frac{J}{m^3}\right)\left(\frac{r}{t^2}\right)\int\frac{dr}{dt}dt \qquad (3.7)$$

$$\frac{1}{q}\frac{\partial(mass \cdot a \cdot r)}{\partial r} = 4\pi q\left(-\frac{1}{m^2}\right)\left(\frac{J}{m^3}\right)\left(\frac{r}{t^2}\right)(r) \qquad (3.8)$$

$$\frac{1}{q}\frac{\partial F}{\partial r} = 4\pi\left(-\frac{q}{m^2}\right)\left(\frac{N \cdot m^2}{m^3 \cdot s^2}\right) \qquad (3.9)$$

$$Charge \equiv Spatial\ Length \cdot Temporal\ Distance \qquad (3.10)$$

$$q_p = l_p \cdot t_p \quad \text{(in Planck units)} \qquad (3.11)$$

28

$$\frac{1}{q}\frac{\partial F}{\partial r} = 4\pi\left(-\frac{q}{m^3}\right)\left(\frac{N\cdot m^2}{m^2\cdot s^2}\right) = 4\pi\left(-\frac{q}{m^3}\right)\left(\frac{N\cdot m^2}{q^2}\right) \qquad (3.12)$$

$$= 4\pi\left(\rho_q\right)\left(\frac{1}{\varepsilon_0}\right)$$

Where $\vec{E}$ is an electric field.

$$\frac{\partial \vec{E}}{\partial r}\vec{a}_r = \frac{\partial \vec{E}_x}{\partial x}\vec{a}_x + \frac{\partial \vec{E}_y}{\partial y}\vec{a}_y + \frac{\partial \vec{E}_z}{\partial z}\vec{a}_z = 4\pi\left(\rho_q\right)\left(\frac{1}{\varepsilon_0}\right) \qquad (3.13)$$

$$\nabla\cdot\vec{E} = \frac{4\pi\rho_q}{\varepsilon_0} \quad \text{(Gauss's Law)} \qquad (3.14)$$

Where "ρ_q" is the proper negative charge density ($-q/m^3$), and "ε_0" is the permittivity of free space.

With hindsight, the eminent physicist Albert Einstein would have been able to derive the six-dimensional electrogravitic EFEs from Gauss' Law. Hindsight is twenty-twenty!

The four-dimensional Einstein EFEs use Einstein's gravitational constant "κ" is the reciprocal of a force which is equal to

$$\kappa = \frac{8\pi G}{c^4} \approx 2.077 \times 10^{-43}\,N^{-1} \qquad (3.15)$$

The six-dimensional EFEs use Mileva's constant "M or μ" for the reciprocal of energy/time or power, in honor of Einstein's first wife, an eminent physicist and active researcher during the conceptual creation and formulation of the Special Theory of Relativity before, during, and after the annus mirabilis papers of 1905. If Albert was the force, Mileva was the energy behind Special Relativity!

$$\mu = \frac{8\pi G}{c^5} \approx 6.923 \times 10^{-52}\,W^{-1} \qquad (3.16)$$

§ 4. How could the gravitational acceleration "g" in six-dimensional space-time be defined by the difference in local scalar curvature near a rogue planet approaching another massive celestial object?

Let us represent the resultant local scalar curvature ξ , with the local scalar curvature tensor $K_{\mu\nu}$ of the gravitational acceleration that would act on a photon traveling between two points, $P_1(x_1, y_1, z_1)$ and $P_2(x_2, y_2, z_2)$, on a metric space where there is significant negative celestial curvature, as the difference between the local scalar curvature tensor $K_{\mu\nu}$ and the celestial curvature tensor, $C_{\mu\nu}$.

$$K_{\mu\nu} - C_{\mu\nu} \equiv \frac{8\pi G}{c^5}\left(E_{\mu\nu} - \hat{\Lambda}_{\mu\nu}\right) \tag{4.1}$$

$$K_{\mu\nu} - C_{\mu\nu} \equiv \frac{4}{\hbar\omega^2}\left(E_{\mu\nu} - \hat{\Lambda}_{\mu\nu}\right) \tag{4.2}$$

The symbol " $\equiv$ " is an equality relating the scalar curvature to the Einsteinian energy density, that is true when values are given to the variables of the equality, such that the mathematical expressions on both sides of the equality symbol produce the same value for all values of the variables within a certain range of validity.

The resultant local scalar curvature is given by

$$\xi_{\mu\nu} \equiv \frac{4}{\hbar\omega^2}\left(E_{\mu\nu} - \hat{\Lambda}_{\mu\nu}\right) \tag{4.3}$$

$$-\xi \equiv \frac{4}{\hbar\omega^2}\left(-\rho_E + p_E\right) \tag{4.4}$$

The resultant gravitational acceleration "g" may be expressed as

$$g = \frac{h\omega^2}{m_{photon}c} = \frac{h\omega^2}{p_{photon}} = \frac{h\omega^2}{\left(\dfrac{h}{\lambda_{photon}}\right)} = \lambda_{photon}\omega^2 \qquad (4.5)$$

$$h\omega^2 \equiv -\frac{8\pi}{\xi}\left(-\rho_E + p_E\right) \qquad (4.6)$$

Thus, the gravitational acceleration may increase significantly as the volume of the remanent energy density decreases as the star goes "spherical supernova", increasing the scalar curvature from the starting curvature ξ_0, and energy is emitted into space-time beyond the original average volume of the star.

$$g \geq -\frac{8\pi}{\left(p_{photon}\right)\xi_0}\left(-\rho_E + p_E\right) \qquad (4.7)$$

$$g \geq -\frac{4\lambda_{photon}}{\hbar\xi_0}\left(-\rho_E + p_E\right) \qquad (4.8)$$

§ 5. Is a dynamic block universe deterministic or indeterministic?

If the present state of the universe is regarded as the causal effect of its unfolding future as the future of times passes through the present and the past in a block universe, it is possible to theorize that a creative intellect that has the ability to interact with that universe at a certain instant may experience the passage of a creator's time in addition to the time of the interaction. Therefore, the creative intellect may know a single equation and its initial values of the block universe to arrive at a certain instant in a deterministic way while also contributing indeterministic change from an inertial frame of reference at an arbitrary point of interaction in the universe. In such scenario, the equation of the universe may be considered both deterministic and indeterministic, or chaotic to a blissfully unaware observer.

§ 6. Is space-time continuous or discrete?

It is possible to think of space-time as a continuous medium at a macroscale for a theory such as the General Theory of Relativity, but space-time may consist of point sources at a Planck scale where space-time emerges eventfully in the dynamic block of space-time. The point sources may be theorized to be interdimensional tunnels or wormholes from the past through the present to the future if the present temporal description is used. In that scenario space-time is porous and, in a sense, discontinuous or discrete since it has nodes of different density between wormholes. Hence, space-time would churn near the Planck length. From that perspective, the smoothness of space-time at a macroscale would be an illusion such as the smoothness of matter is an illusion to our sight even though we understand that matter is very porous. There is more spatial length between the atoms of matter than across the boundaries of the mass of any of its fundamental particles. Consequently, continuity or discreteness coexist in nature providing space-time that is emergent at the fundamental level and smooth and stable at the level of the manifestations of mass or matter. Space-time is transcendental in nature, while the foundation of our mathematics is discrete for the benefit of our intellect. (Taylor, 1966)

§ 7. The information paradox for a black hole

It is possible to speculate that when a particle falls into a black hole the information on the particle is converted into gravitational waves that is accessible outside of the blackhole. The information may be about the wavelength, amplitude, electric charge, pressure, composition, spin, density, temperature, shape, size, mass, etc., of each particle. The information is encoded in the correlations between future gravitational waves and past gravitational waves that emerge from the black hole as the particle is swallowed by the black hole. Even though the particle is out of sight to an observer beyond the event horizon of the black hole, its informational footprint stays in its universe and it is imprinted in the surrounding spacetime. The information footprint of the particle becomes part of the spatiotemporal disturbance of the black hole. Hence, the information is always conserved. However, it is believed that the remaining black hole can be described by only its mass, spin, and electric charge. (Melia, 2007)

Spacetime is believed to be full of particle-antiparticles pairs that emerge into existence due to correlated quantum effects. Each particle that falls into the black hole carries negative energy inward and breaks its correlation with its partner particle outside. So, the black hole steadily loses mass. As a consequence, if the black hole does not feed on any ordinary matter, then it would eventually evaporate.

Thus, the entire history of the position of each and every single particle existing in the universe, before or after, the particle falls into a black hole, can be retraced back to the point singularity. (Wald, 1977) In this way, quantum determinism, and time reversible symmetry, would still predict the future evolution of a particle if you have access to the particle's history. The quantum gravitational waves of a particle that falls into a black hole merges General Relativity, or Classical Physics, with Quantum Mechanics, or Quantum Physics. The information is preserved by undulation(s), or wavelet(s), on the surface, or layer, of the event horizon which would record the information from the radiating particle near the black hole.

PART II

Chapter 3

Pressure equals Energy Density: A Natural Law

§ 1. What is a field of force?

A field of force is a region of space-time where a quantity may be
measured at any spatiotemporal point in the field. The field can be
felt as a force by elementary particles, or objects submerged in the
field. The field of force is the gradient of the field potential state of
the spatiotemporal wave medium at any given instant of time.
The Kalachakra Mandala, or the wheel of time, may be respectfully
considered as a good depiction of the spatiotemporal field, as any
point in space-time may expand or contract, in a cyclical
spatiotemporal wave medium.

If two ideal mirrors face each other with a totally reflecting surface
material, the reflections between the two mirrors would have no
limit, going to infinite reflections as time goes to eternity. If an
observer were to look at one of the mirrors, the view resembles the
depiction of the Kalachakra Mandala, the expanding space from a
single spatiotemporal source at inner eternity from the past to the
future of time. The effect of the reflections renders a view of the
light waves as if the light waves were traveling through space-time
from the sink to the point source of the expansion of space-time, or
from real to imaginary space-time.

The present strength of the classical electromagnetic force in
comparison to the strength of the classical gravitational force is
greater by orders of magnitude. However, at the beginning of the
universe before and after the big bang event the strength of the
gravitational force has gradually weakened due to the expansion of
space-time, a farther field, and the distribution and manifestation of
light and mass throughout the universe while the electromagnetic
force has remained largely a more localized and quantized force at
more condensed distributions of charge, a near field, or in the
structures of mass of the present universe.

As stated by Paul Dirac at the Lindau Conference in 1979 on his large numbers hypothesis, the black body radiation of the cosmic microwave background, the frequency and the temperature decrease, the wavelength of the components and the expansion of the distance to the galaxies increase, according to the law of time by using $\sqrt[3]{t}$. Is it possible that Dirac was arriving at conclusions of his Einstein-De Sitter theory of large numbers due to his unknowingly acceptance of the three dimensionality of time?

The Casimir effect force "F" may be considered a force per unit of area "S" produced by negative energy, that may come from virtual particles inside a spatiotemporal bridge (wormhole) similar to a source of magnetic field within a traversable spatiotemporal bridge, pushing onto its walls. Such wormhole may also allow quantum entanglement between particles.

$$\frac{\partial F}{\partial S} = -\frac{\pi^2 hc}{240 L^4} \qquad (1.1)$$

Today, theoretical physicists and researchers are traversing spatiotemporal bridges with their hypothetical equations, but eventually, as scientific knowledge progresses, it will be astronauts and explorers in their spaceships.

In Quantum Mechanics, the entanglement of the states of two particles is a kind of superposition, but not every superposition of the states of two particles is entangled.

Do we know if gravity exists at a particle level? We have not been able to detect gravitational effects at the particle level. Thus, is it not possible that gravity does not apply at all on that scale? We know that gravity acts at the particle level and that it adds up to the gravity that we experience at a classical level. A gravitational theory that explains how gravity is able to act from the Planck scale (or Quantum scale) to the classical scale of the General Theory of Relativity is needed. At what level gravity and the three fundamental forces become an emergent phenomena? Is the composition of space-time also an emergent phenomenon? The dynamic theory of space-time provides an explanation and an answer to these questions.

§ 2. The photonic field

The structure of the photonic field is usually represented by two components, the electric field, and the magnetic field, at quadrature. However, the photonic field may be represented by a third field which may be considered a triadic electromagnetic field. The photon has been described as a quantized bundle of electromagnetic energy that acts like a particle, with no rest mass, traveling at the speed of light, which has momentum. The quantized energy of the photon is given by the Planck constant times the frequency of the photon.

Let us imagine a corpuscle as the basic constituent of light, as the corpuscle, or photon, travels through the spatiotemporal medium after it was emitted from a light source such as the sun. The trajectory of the photon, or the photonic field, may be represented as the resultant trajectory of the electric field component and the magnetic field component. The electric field travels on a temporal plane and the magnetic field travels on spatial plane that may be represented at quadrature. The electric field is the trace, or reflection, of the photonic field on the temporal plane and the magnetic field is the trace of the photonic field on the spatial plane. Each trace shows how the photonic field changes over time, or over space, as it propagates through spacetime. So, the structure may be considered a triadic electromagnetic field.

The photonic field travels at a small phase angle from the electric field since the electric field may be orders of magnitude larger than the magnetic field as in $E = vB$, where v is the speed of light. It is hypothesized that this phase angle is the result of the relativistic mass of the photon as it travels through spacetime. If the polarization of the photon is counterclockwise in the direction of propagation, the phase angle of the photon is lagging due to its relativistic mass. Thus, the amplitude of the photonic field is closer to the amplitude of the electric field in this case, regardless of handedness of polarization.

The momentum associated with the photonic field is also associated with each electric and magnetic component. The radial kinetic vector of the photonic field may also be represented by the electric and magnetic vectors perpendicular to the direction of propagation. The kinetic vector may be augmented and continually directed as a force.

Let us now imagine a coil of fiber optic cable connected to a laser source that emits an optical signal that is optically rectified through a nonlinear process and filtered, and then propagates through the entire length of the cable. The electric component can be vertically filtered, and the magnetic component can be horizontally filtered. Hence, the swift oscillations of both components are optically rectified and only the envelope of the optical signal would remain.

Both components would be a quasi-DC nonlinear polarization. The production of terahertz pulses and radiation is possible as a second order effect in non-absorbing electro-optical material. The polarization may be commutated at very high speed without limit. The polarized optical signal would travel through spacetime with a lower phase velocity due to the phononic oscillations of the atomic lattices in the non-absorbing electro-optical material that would possibly radiate in a Cerenkov cone arrangement.

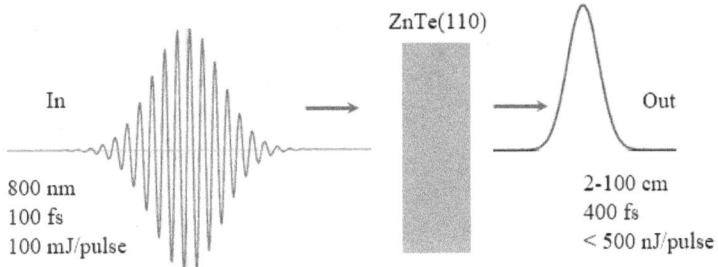

Figure 1. Optical or Laser Resonator

Let us consider the propagation of light in an optical resonator through a transparent homogeneous medium with a mode of operation and a reflective mirror at its end. Standing waves are created when light waves traveling back and forth through the medium interfere with each other. Only light waves whose round-trip distance is integer multiples of the fundamental wavelength will become a standing wave.

The condition of the light is self-consistent, a mode reproduces its transverse amplitude after a full resonator round trip. During the return trip the mode profile may change in size and shape, while preserving its optical rectification. The optical phase is reproduced after one trip as a multiple of 2π, and the overall optical power may be attenuated.

The resonator modes exist for the resonant (optical) frequencies. The phase shift after a round trip depends on the intensity pattern of a mode. A laser resonator produces significant power in single mode operation, only a single mode is excited. In a Gaussian mode, the laser resonator would have an ideal beam quality, the output would have limited diffraction. Both the single mode of operation and the Gaussian mode of fiber have similar shapes. A single-mode fiber guarantees a fixed intensity profile at its output, assuming that all light launched into cladding modes, or unguided modes, is lost before the fiber end is reached.

It would be interesting to test such a coil to determine and optimize the number of turns, type of fiber optic cable, physical size, etc., necessary to measure the directional resultant force in a gravitational field or in free space. Multiple coils may be connected in series or parallel to modify, optimize, and direct the resultant force of the photonic field. The optical fibers may dissipate heat, and produce sound, vibration, and gravitational waves, from the laser mode oscillations in the optical medium of the optical resonator. This optical rectification technology may have the potential to be a novel propulsion technology for a photonic field drive. The elementary bosons, such as photons, gluons, W and Z, are force carriers that function as the glue holding matter together. Could gluons, W, or Z bosons be used to create bosonic field drive technologies? What kind of a compact power source and a particle accelerator may be used for a bosonic field drive?

How would a brilliant aeronautics engineer like Juan de la Cierva use this propulsion technology?

Let us now use our imagination to propose a fantastic spacecraft that Juan de la Cierva y Codorniu, as the father of the helicopter in the early 20th century, called an autogyro, may have dreamed about as a young man inspired by the science-fiction ideas of E. Gaspar, Jules Verne, and K. Tsiolkovsky, for diversion when not studying for his civil engineering exams at the university. The spacecraft would be 40 to 50 feet long, with an ellipsoidal tic-tac shape, and four telescoping girder-like landing legs. The spacecraft would have a plug door, designed to seal itself by taking advantage of the pressure difference on its two sides, towards the middle right side of the fuselage, a windowless cockpit with an external visibility system that connects to a wraparound display, side portholes that are slim and oblong, a pilot and a co-pilot, and white paint on the fuselage to reflect sunlight reducing heat and radiation. The spatiotemporal curvature of the bosonic field drive would serve as a significant defensive shield, a spatiotemporal streamlining for faster speed, and as an effective cloaking device. Sunlight around the spacecraft would be diffused by the bosonic field, as the sunlight bounces off to travel through the spatiotemporal medium that changes its angles and scatters it in all directions. Sunlight would seem to wrap around the spacecraft in a softer and fuzzy way that would not cast any harsh shadows.

The spacecraft would have the ability to generate sufficient lift to sustain level flight, hover, climb and descent, like an autogyro. An additional liquid fueled Goddard rocket engine could be installed under the center of the fuselage for rapid takeoff until the bosonic field drive was fully operational. The bosonic drive would allow the spacecraft to make rapid lateral movements, steep vertical climbs, or rapid descents, at an incredible rate of speed without sonic booms. The bosonic field would envelope the spacecraft and would extend for quite a distance, a pilot would be able to see the frothy water, or white water disturbance effect on the ocean water below, if the spacecraft flew at a consistent low altitude making lateral moves, above the surface of the ocean.

The bosonic field drive would consist of multiple fiber optic coils around the geometry of the spacecraft that could work in tandem to balance, assist, or counteract, the unbalanced forces around the vehicle. These coils may be connected to a backbone loop to serve as a return and as a bosonic afterburner. This technology would be a lot different than developing a direct rotor control on an autogyro through cyclic pitch variation, but to a very capable inventor like Juan de la Cierva, it was all a matter of engineering. The fiber optic coils may be stacked for greater lift or decent, installed concentric or staggered in an inner, middle, and outer hull, to facilitate effective cooling and ventilation, radiation shielding, and connected in series for balance during hover or level flight. The bosonic field drive may produce enough gravitational acceleration inside the vehicle using wave guides and resonators and outside the outer hull to dampen, or cancel, inertia during rapid acceleration or deceleration. A very reliable heat dissipation system would be necessary for the coils not to overheat due to particle decay or impedance at full power. When cruising through space at full speed, the fiber optic coils of the propulsion system may be powered down for cooldown and energy conservation. The control system for the craft would require a very high speed Turing machine (computer) and commutator (switching mechanism), and a life support system to operate in free space.

What device could possibly emit light through a process of optical amplification based on the stimulated emission of electromagnetic radiation? What kind of a rechargeable and very compact power source could emit bosons that can be propelled through a particle accelerator? Juan de la Cierva being a very practical problem solver knew that perhaps Tesla technology (a compact turbine, an optical device, a compact generator) might provide solutions. Today, there are lasers, particle accelerators, computers, fiber optic cables, and the related technology, that may help an inventor like Juan de la Cierva to fulfill such a science fiction dream.

Cierva's remaining questions would have probably been: Was Nikola Tesla already working on a bosonic field drive? Could a bosonic propulsion system be built in the 20th century?

§ 3. Why is acceleration represented by a unit of spatial distance per every two units of temporal distance?

Every spatial dimension has an orthogonal temporal dimension. What happens to the orthogonal temporal dimensions of the other two spatial dimensions? Will they not be in the way of a different spatial dimension or parallel to it?

A one-dimensional spatial segment along one of the spatial rectangular coordinate axes may have temporal extensions that are related to other spatial axes, since time is three-dimensional like space. For instance, the temporal distance element of the Friedman-Lemaitre-Robertson-Walker metric for space-time is given in squared seconds just like the spatial distance element is given in squared meters. Time is non-linear. The existence of proper time proves that time is non-linear like space. Time endows space, then space endows more time. For a linear one-dimensional spatial

40

segment there are two ends, each end of the segment represents a direction where space may endow time to emerge. Consequently, for any spatial segment there may be two units of temporal distance emerging at both ends. Acceleration is the dimensional ratio of the spatial distance traveled by an object through a spatial segment with the temporal distances that emerge at both ends of the segment of the trajectory, m/s^2.

On a spatiotemporal wave, the radian trajectory is the spatiotemporal distance traveled by a particle while the displacement is half of the period during a half cycle. In this case, the displacement is identified as the translation that maps the initial position to the final position of the particle. The spatiotemporal direction of the displacement may not align with the temporal coordinate direction, or the spatial coordinate direction, of a rectangular coordinate system, for the path. *The spatiotemporal distance of the radian trajectory "C" is not relativistic, but the displacement, the coordinate spatial distance "s_x" of the amplitude, the coordinate temporal distance "t_x" of the period, of the spatiotemporal wave, are relativistic and reciprocal.*

Let us represent this reciprocal relationship mathematically.

$$C = 2\pi \cdot \sqrt{\frac{s_x^2 + t_x^2}{2}} \qquad (3.1)$$

$$\frac{2C^2}{4\pi^2} = \frac{C^2}{2\pi^2} = \frac{\left(\frac{C}{\sqrt{2}}\right)^2}{\pi^2} = s_x^2 + t_x^2 \qquad (3.2)$$

$$\pi = \frac{\frac{C}{\sqrt{2}}}{\sqrt{s_x^2 + t_x^2}} \qquad (3.3)$$

Hence, the effective value of the circumference of the spatiotemporal wave or wave function divided by the square root of the sum of the squares (a segment) of the spatial length (semi-major axis) and the temporal distance (semi-minor axis) is equal to the transcendental

41

number π or $\left[\left(-\frac{1}{2}\right)!\right]^2$. The circumference is absolute space divided by relativistic space-time resulting in the transcendental ratio π.

On a Euclidean plane, the radian trajectory "C" of a standing spatiotemporal wave provides the transcendental value of π given by

$$\pi = \cfrac{C}{\left(t_x + s_x\right) \sum_{n=0}^{\infty} \binom{0.5}{n}^2 \left[\frac{\left(t_x - s_x\right)^2}{\left(t_x + s_x\right)^2}\right]^n} \tag{3.4}$$

$$\pi = \cfrac{C}{\left(t_x + s_x\right)\left(1 + \frac{1}{4}\left[\frac{\left(t_x - s_x\right)^2}{\left(t_x + s_x\right)^2}\right] + \frac{1}{64}\left[\frac{\left(t_x - s_x\right)^2}{\left(t_x + s_x\right)^2}\right]^2 + \frac{1}{256}\left[\frac{\left(t_x - s_x\right)^2}{\left(t_x + s_x\right)^2}\right]^3 + ...\right)} \tag{3.5}$$

Consequently, from previous research, the transcendental number π may be represented by an acceleration of an area, but it represents the emergent relativistic ratio of the topological geometry of a spatiotemporal wave function at a quantum level. (Nieves, 2020) Therefore, the transcendental number π is relativistic in its geometrical spatiotemporal nature.

It is interesting to point out that the sum of the angles inside a triangle on a Euclidean plane equals π. The total curvature of a geodesic triangle equals the deviation of the sum of its angles from π. The attributes of the Special or the General Theory of Relativity spring from a Quantum Theory of Quantum Mechanics. The quantum relativistic attribute of the spatiotemporal wave function scales up to the classical world arounds us in our universe.

In its simplest definition, curvature is the numerical quantity by which a curve deviates from a straight line, or a curved surface deviates from a Euclidean plane. The curvature of a space, or for a surface, which is locally isotropic and homogeneous is described by a single Gaussian curvature. Curvature may be defined as the numerical quantity by which a Gaussian spatiotemporal surface deviates from being a Euclidean spatiotemporal plane.

A Gaussian curvature can be defined without reference to an embedding space, such an intrinsically curved two-dimensional spatial surface is a simple example of a Riemannian manifold.

The Ricci scalar, or the scalar curvature, is the simplest curvature invariant of a Riemannian manifold. The scalar curvature is given by a single real number for each point of a Riemannian manifold determined by the intrinsic geometry near the point.

The scalar curvature may be described by the amount of change in the volume of a small geodesic ball in a Riemannian manifold from the same ball in a Euclidean space.

Let us imagine a variable triangle in a metric space as a set of points in conjunction with a metric on the set. The function of the metric defines a distance between any two points of the set. The metric satisfies the topological properties of the spatiotemporal Gaussian surface.

The curvature of the Gaussian spatiotemporal surface is an intrinsic measure that only depends on distances that are measured on the surface itself.

The variable triangular surface moves in such a way that the sum of the reciprocals of its intercepts on the three Cartesian coordinate axes is a constant *"d"*.

The interior angle of any one of its three corners is equal to the arc sine of two times the area of the triangle divided by the product of the lengths of the segments of the two sides of the corner that has the interior angle.

$$\theta_x = Arc\,Sin\left(\frac{2 \cdot Area}{\overline{ac} \cdot \overline{ab}}\right) \qquad (3.6)$$

Where *a, b,* and *c* are the intercepts on the *x, y,* and *z* axes, respectively.

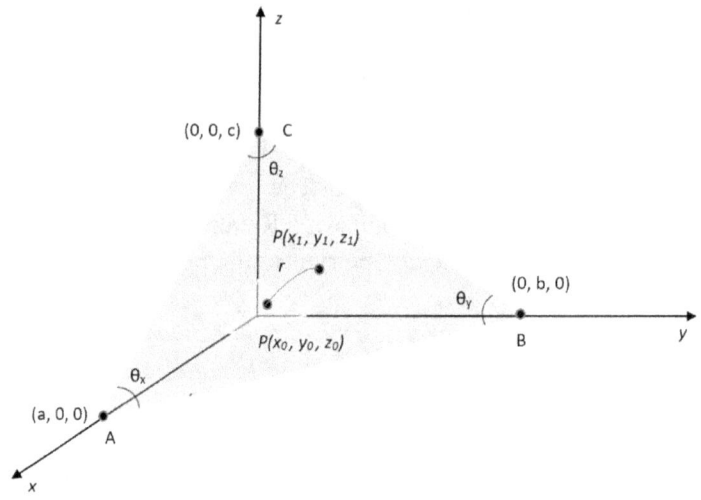

Figure 2. A variable triangular surface

Since the sum of the reciprocals of the intercepts on the Cartesian coordinate axes is equal to a constant "*d*", we have

$$\frac{1}{a}+\frac{1}{b}+\frac{1}{c}=d \qquad (3.7)$$

where "*d*" is a constant whose equation can be denoted as

$$\frac{1}{a}\left(\frac{1}{d}\right)+\frac{1}{b}\left(\frac{1}{d}\right)+\frac{1}{c}\left(\frac{1}{d}\right)=1 \qquad (3.8)$$

The above equation indicates that the triangular surface passes through the fixed point $P\left(\dfrac{1}{d},\dfrac{1}{d},\dfrac{1}{d}\right)$.

The triangular surface meets the Cartesian coordinate axes at *A*, *B*, and *C*, respectively, so that the intercepts on the *x*, *y*, and *z* axes are: $a = OA$, $b = OB$, and $c = OC$. Let us denote the triangular surface as $\vec{r} \cdot \vec{p} = d$.

The spatial distance vectors at points *A*, *B*, and *C*, are: $a\vec{e}_x$, $b\vec{e}_y$, $c\vec{e}_z$.

44

Since the points *A, B,* and *C* lie on the triangular surface, we obtain

$$\vec{e}_x \cdot \vec{p} = \frac{d}{a} \tag{3.9}$$

$$\vec{e}_y \cdot \vec{p} = \frac{d}{b} \tag{3.10}$$

$$\vec{e}_z \cdot \vec{p} = \frac{d}{c} \tag{3.11}$$

Thus, let us substitute for $\vec{r}$ to arrive at the intercept form of the variable triangular surface,

$$\vec{r} = x\vec{e}_x + y\vec{e}_y + z\vec{e}_z \tag{3.12}$$

$$\vec{r} \cdot \vec{p} = x\vec{e}_x \cdot \vec{p} + y\vec{e}_y \cdot \vec{p} + z\vec{e}_z \cdot \vec{p} = d \tag{3.13}$$

$$x\left(\frac{d}{a}\right) + y\left(\frac{d}{b}\right) + z\left(\frac{d}{c}\right) = d \tag{3.14}$$

Consequently, the intercept form of the variable triangular surface equation is given by

$$\frac{x}{a} + \frac{y}{b} + \frac{z}{c} = 1 \tag{3.15}$$

It is interesting to note that the altitudes of an equilateral variable triangular surface intercept at the orthocenter. The altitude, perpendicular bisector, angle bisector, and median from the vertex angle to the base of an equilateral triangle are all the same segments.

The total curvature of a geodesic triangle on a Gaussian surface equals the deviation of the sum of its interior angles from π. If a surface has positive total curvature, the sum of the angles of a geodesic triangle on that surface will exceed π. If it has negative total curvature, the sum of the interior angles will be less than π. If a triangle is on a Euclidean plane, which is a flat surface with zero total curvature, the interior angles will sum to precisely π radians.

For higher-dimensional manifolds and surfaces that are embedded in a Euclidean space, the concept of curvature is complex, and it depends on the chosen direction for the manifold or surface.

The concept of radial scalar curvature relies on the ability to compare a curved space with another space that has zero curvature, or a constant curvature like a sector on a sphere.

Let us define the radial scalar curvature operator as:

The spatial radial curvature operator,

$$\Gamma_{\nabla}^2 = \pi^2 \cdot \nabla^2 = \pi^2 \left(\frac{\partial^2}{\partial x_0^2} + \frac{\partial^2}{\partial y_0^2} + \frac{\partial^2}{\partial z_0^2} \right) \tag{3.16}$$

The temporal radial curvature operator,

$$\Gamma_{\odot}^2 = \frac{\pi^2}{c^2} \cdot \odot^2 = \frac{\pi^2}{c^2} \left(\frac{\partial^2}{\partial t_{x_0}^2} + \frac{\partial^2}{\partial t_{y_0}^2} + \frac{\partial^2}{\partial t_{z_0}^2} \right) \tag{3.17}$$

The spatiotemporal radial curvature operator,

$$\Gamma_{SP}^2 = \Gamma_{\nabla}^2 + \Gamma_{\odot}^2 \tag{3.18}$$

Applying the operator on the complex wave function,

$$\Gamma_{SP}^2 \Psi(r,t) = \Gamma_{\nabla}^2 \Psi(r) + \Gamma_{\odot}^2 \Psi(t) \tag{3.19}$$

Let us now describe mathematically how it is possible to use a triangle as the radial measure of scalar curvature at an arbitrary quantum point or at a classical point from an origin of zero curvature with interior angles inside a triangle equal to θ_0 in a region of space-time.

Let us imagine a triangle on a spatiotemporal manifold that has no curvature with each corner on each of the positive axes x, y, and z.

46

Then, let us also imagine a second triangle farther away on a spatiotemporal manifold with curvature, with each corner on each of the positive axes x, y, and z. The distance "r" of a proper spatial segment from the orthocenter point of the first triangle $P(x_0, y_0, z_0)$ to the orthocenter point of the second triangle $P(x_1, y_1, z_1)$ represents a radial distance of scalar curvature. Each interior angle of the triangle would be identified by the closest axis to it.

$$\vec{r} = |r| \angle \phi = |r| e^{\sqrt{\pi}(\phi)} \vec{a}_r \tag{3.20}$$

For the positive axes chosen, we have

$$r = (r_1 - r_0) = \sqrt[2]{(x_1 - x_0)^2 + (y_1 - y_0)^2 + (z_1 - z_0)^2} \tag{3.21}$$

$$\tau = (t_1 - t_0) = \sqrt[2]{\left(t_{x_1} - t_{x_0}\right)^2 + \left(t_{y_1} - t_{y_0}\right)^2 + \left(t_{z_1} - t_{z_0}\right)^2} \tag{3.22}$$

$$\phi = (\theta_1 - \theta_0) + (\theta_2 - \theta_0) + (\theta_3 - \theta_0) = \theta_x + \theta_y + \theta_z \tag{3.23}$$

Curvature can be defined by the square of the derivative of the angle θ of a sector with respect to the length of its arc S.

$$S = r\theta \quad \text{(A Sector)} \tag{3.24}$$

$$\frac{1}{r^2} = \left(\frac{\partial \theta}{\partial S}\right)^2 \equiv \left(\frac{change\ in\ angle\ \theta}{change\ in\ radians}\right)^2 \tag{3.25}$$

Therefore, we may describe curvature in a spatiotemporal wave as the square of the ratio of the change in the temporal coordinate distance with the change in the radian trajectory distance, to yield the reciprocal of the area of the surface of curvature. It is interesting to observe that the surface of curvature emerges as a temporal surface to manifest its reciprocal as a spatial curvature.

For the radial scalar curvature of a triangle when the change in radians equals π,

$$R = \frac{\partial^2\left(-[\Psi(r)]^2\right)}{\partial r^2} = \frac{\partial^2\left\{\left(\frac{1}{\ln e^{-r}}\right)\right\}}{\partial r^2} \tag{3.26}$$

$$= \frac{\partial^2\left[-\frac{1}{r}\right]}{\partial r^2} = \frac{1}{r^2}$$

$$R = \frac{\partial^2\left(-[\Psi(r)]^2\right)}{\partial r^2} \tag{3.27}$$

$$= \left(\frac{\partial(\theta_1 - \theta_0)}{\partial S}\right)^2 + \left(\frac{\partial(\theta_2 - \theta_0)}{\partial S}\right)^2 + \left(\frac{\partial(\theta_3 - \theta_0)}{\partial S}\right)^2$$

$$= \frac{\partial\theta_x^2}{\pi^2} + \frac{\partial\theta_y^2}{\pi^2} + \frac{\partial\theta_z^2}{\pi^2}$$

$$\pi^2\left(\frac{\partial^2\Psi_x^2}{\partial x_0^2} + \frac{\partial^2\Psi_y^2}{\partial y_0^2} + \frac{\partial^2\Psi_z^2}{\partial z_0^2}\right) = \partial\theta_x^2 + \partial\theta_y^2 + \partial\theta_z^2 \tag{3.28}$$

$$R = \frac{\partial^2\left(-[\Psi(ct)]^2\right)}{c^2\partial t^2} = \frac{\partial\theta_{t_{x0}}^2}{\pi^2} + \frac{\partial\theta_{t_{y0}}^2}{\pi^2} + \frac{\partial\theta_{t_{z0}}^2}{\pi^2} \tag{3.29}$$

$$\pi^2\left(\frac{\partial^2\Psi_{t_x}^2}{\partial t_{x0}^2} + \frac{\partial^2\Psi_{t_y}^2}{\partial t_{y0}^2} + \frac{\partial^2\Psi_{t_z}^2}{\partial t_{z0}^2}\right) = \partial\theta_{t_{x0}}^2 + \partial\theta_{t_{y0}}^2 + \partial\theta_{t_{z0}}^2 \tag{3.30}$$

Therefore, the radial curvature is reciprocal to the sum of the squares of the changes in the interior angles of the triangle. Radial scalar curvature increases if the sum of the changes of the interior angles is positive, positive curvature increases. Conversely, radial scalar curvature decreases if the sum of the changes of the interior angles is negative, negative curvature increases, as the interior angles are compared to the interior angles of the triangle at zero curvature in flat space.

48

Where R is a real number representing the trace of the radial scalar curvature at a point $p(x, y, z)$ on the surface of a triangle, "r" is the radial distance, "τ" is proper time, π is a transcendental number, $\odot^2$ is the double tempus operator, $\Psi(r)$ is the spatial wave function equal to $1/\sqrt{r}$, but the second derivative of the negative or temporal squared wave function with respect to "r", is the radial scalar curvature of space, $d^2\left(-\left[\Psi(r)\right]^2\right)/dr^2 = 1/r^2$, and θ_n is an interior angle of a triangle that may or may not be geodesic.

Hence, *the wave function represents the curvature of space-time as a resultant wave produced by the interference of the spatiotemporal wavelets at every spatiotemporal point*. The more energy (fields or particles) that is present in the single wave function, the greater the frequency of the wave function, and the greater the curvature. In the total quantum system of multiple particles, even though each particle has its own wave function, there is a single total wave function representing all the particles.

The total wavefunction consist of the spatial wavefunction and the spin. Two photons (two bosons), who share the same quantum state, always have a total symmetric wave function, the wavefunctions reinforce each other as the surface wavefunction oscillates. On the other hand, two electrons, who do not share the same quantum state, always have an anti-symmetric total wave function, their wavefunctions would offset each other and their probability would be zero.

With the exclusion of all other forces between the particles, the total wave function would either manifest a spatiotemporal symmetric attraction or anti-symmetric repulsion, in the geometry of the total wave function, that depends on the wave functions and the spins of the particles involved. Electrons, and other fermions, may have symmetric wavefunctions at their lowest energy level, even if their spins are anti-symmetric. Their wavefunctions of the electrons may alternate in unison, but the spins may alternate out of phase. Aside, at very low temperatures, an electron pair may become a Cooper pair, or a boson of superconductivity.

The condition of the two electrons (two fermions), of not sharing the same quantum state, is what manifests the Pauli exclusion principle, which is the assertion that no two fermions can have the same quantum number. The Pauli exclusion principle occurs when the wavefunctions are out of phase and offset each other. The Pauli exclusion principle prevents all the electrons in an atom from falling at the same time to their lowest energy level.

It is interesting to point out that the emerging wave function is temporal in nature, with a negative or positive linear curvature, and able to produce non-linear curvature for either time or space.

$$\Psi(r) = \sqrt{-\frac{1}{\ln\ e^{-r}}} = i\sqrt{\frac{1}{\ln\ e^{-ct}}} = \frac{1}{\sqrt{r}} = \frac{1}{\sqrt{ct}} \qquad (3.31)$$

Space endows time, then time emerges, which endows more space.

Let us consider a graph of a rectangular coordinate system of the one-dimensional wave function, $\Psi(r)$, with an axis for "r" (the x-axis), let us call it the r-axis, and an axis for the wave function, (the y-axis). A value of "r" in the one-dimensional wave function renders a value of the linear curvature, $1/r$, at any point on the curve. Linear curvature is the reciprocal of the exponent of the kernel of natural growth, $e^{\pm r}$. A value of "ct" may be substituted as a temporal equivalent for "r" to analyze the one-dimensional temporal wave function. The vertical distance from any value on the r-axis to a point on the curve is identified as the probability density $|\Psi(r)|^2$ of finding a point particle on the wave at any given instant of time. Hence, a one-dimensional wave function graph is a "linear curvature versus spatial distance" graph that is helpful to visualize the probability density of the wave function along the entire trajectory of the wave. The steeper the curvature of the curve within a specific range of "r", the larger the probability to find the particle in that distance of "r", as the particle follows its trajectory. So, the probability density is higher, because the length of the trajectory is greater in the range of "r" where the curvature is steeper.

A wave function may also be two-dimensional, $-\left[\Psi(r)\right]^2$, with non-linear curvature, $1/r^2$, for the radial scalar curvature. Non-linear curvature is the reciprocal of the squared exponent of the kernel of natural growth, $e^{\pm r}$. Furthermore, it is interesting to observe that either linear or non-linear curvature is related to the kernel of natural growth which is a wave. Therefore, *curvature is produced by the interference of spatiotemporal wave functions*. The distance "r" is a function of the three-dimensional space (x, y, and z) as previously described. A three-dimensional rectangular coordinate system graph with coordinate axes for x, y, and z, for the two-dimensional wave function would render the probable location of the point particle on the surface graph. The distance between a point on the x-y plane and a point on the surface graph yields the probability density $\left|-\left[\Psi(r)\right]^2\right|^2 = \left|\Psi(r)\right|^4$ of finding the point particle on the surface at any given instant of time. Thus, a two-dimensional wave function graph is a "non-linear curvature versus spatial surface area" graph to visualize the probability density of the wave function on the entire surface of the wave. The probability density is higher, because the volume of the possible trajectories is greater in the surface area "r^2" where the non-linear curvature is steeper.

§ 4. Could spatiotemporal curvature be represented as probability? How can the Einstein field equations be represented by a wavefunction?

Let us consider a tesseract, which has an inner spatiotemporal cube that is half of the expanded outer spatiotemporal cube.

For the volume of an inner spatiotemporal cube, the partial derivative of the volume with respect to space-time is given by

$$\frac{\partial s^{n-1}}{\partial s} = (n-1)s^{n-2} \tag{4.1}$$

$$s^{n-2} = \frac{1}{(n-1)}\frac{\partial s^{n-1}}{\partial s} \tag{4.2}$$

The derivative of the spatiotemporal volume illustrates the probabilities of the three surfaces on the cube that are parallel to the three Cartesian coordinate planes. For four dimensions, the outer spatiotemporal cube is folded, so the inner cube becomes a three-dimensional spatial volume and time is treated as non-linear, even though it is regarded as one-dimensional. This is a misconception in the original EFEs, including the fraction of "½" for static space that should be "⅓" as calculated for four-dimensional space-time. In mathematics, a fudge factor may be described as a term inserted into a formula to allow for an uncertainty, or to make something congruent with an expected or desired result.

There are two unintentional, or bona fide fudge factors in the original EFEs, the value of 6 for the contracted metric tensor, $g^{\mu\nu} g_{\mu\nu} = 6$, which if space-time is four-dimensional, it should be 4, and the tripling of the energy density "3ρ" in the four-dimensional matrix of the stress-energy-momentum tensor to balance the folded time-time component "ρ" with the three space-space components of pressure "p".

However, for six dimensions, three spatial and three temporal dimensions, the equation emerges beautifully and natural to the complex geometry of spacetime, the fraction "⅕" accounts for the unfound ordinary and baryonic matter and energy of the universe, the energy density and pressure of the stress-energy-momentum tensor have balanced coefficients, and curvature is to the fourth power to represent the wavefunction, without folding time. Therefore, spatiotemporal curvature can be represented as probability more accurately.

If $n = 6$, for six dimensions, we get

$$\frac{\partial s^5}{\partial s} = 5s^4 \tag{4.3}$$

$$s^4 = \frac{1}{5} \frac{\partial s^5}{\partial s} \tag{4.4}$$

If $n = 4$, for four dimensions we obtain

$$\frac{\partial s^3}{\partial s} = 3s^2 \tag{4.5}$$

$$s^2 = \frac{1}{3}\frac{\partial s^3}{\partial s} \tag{4.6}$$

The inner spatiotemporal cube may be denoted as

$$s_i^2 = \frac{r_i^2 \cdot t_i^2}{2} \tag{4.7}$$

$$r_i^2 \cdot t_i^2 = \frac{2}{3}\frac{\partial \left(\left\|\Psi_i(r,t)\right\|\right)^{-3}}{\partial \left(\left\|\Psi_i(r,t)\right\|\right)} = \frac{1}{3}\left[(2)\cdot(-3)\cdot\left(\left\|\Psi_i(r,t)\right\|\right)^{-4}\right] \tag{4.8}$$

$$= \frac{1}{3}\left(\frac{-6}{\left(\left\|\Psi_i(r,t)\right\|\right)^4}\right)$$

The outer spatiotemporal cube may be expressed as

$$s_o^2 = r_o^2 \cdot t_o^2 = \frac{1}{\left(\left\|\Psi_o(r,t)\right\|\right)^4} \tag{4.9}$$

From the four-dimensional Lorentzian metric tensor, we divide the six diagonal components of the Ricci tensor by the same diagonal components of the metric tensor, to obtain

$$g^{\mu\nu}g_{\mu\nu} = g = 6 \tag{4.10}$$

General Relativity reduces to Special Relativity in sufficiently flat spatiotemporal regions. Non-flat or nearly-flat "$g_{\mu\nu}$" may be reduced to "$\eta_{\mu\nu}$" (flat Minkowski metric) in sufficiently small spatiotemporal regions. The flat Minkowski metric is the solution for zero curvature in the EFEs.

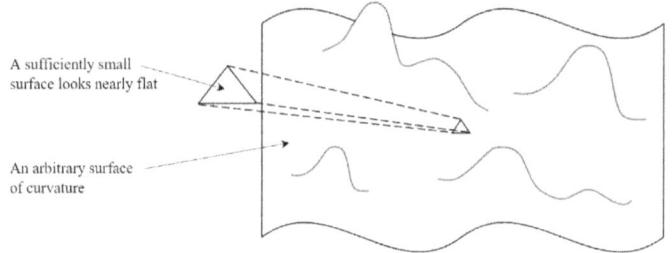

A sufficiently small surface looks nearly flat

An arbitrary surface of curvature

Figure 3. A Nearly Flat Spatiotemporal Region

The difference between the outer and inner spatiotemporal cubes is

$$\frac{1}{\left(\left|\Psi_o(r,t)\right|\right)^4} - \frac{1}{3}(6)\left(\frac{-6}{\left(\left|\Psi_i(r,t)\right|\right)^4}\right) \qquad (4.11)$$

Expressing the wave function in terms of the Ricci curvature,

$$R_{\mu\nu} - \frac{1}{3}g_{\mu\nu}R = \frac{1}{\left(\left|\Psi_o(r,t)\right|\right)^4} - \frac{1}{3}(6)\left(\frac{-6}{\left(\left|\Psi_i(r,t)\right|\right)^4}\right) \qquad (4.12)$$

Assembling the rest of the EFEs, we get

$$\frac{1}{\left(\left|\Psi_o(r,t)\right|\right)^4} - \frac{1}{3}(6)\left(\frac{-6}{\left(\left|\Psi_i(r,t)\right|\right)^4}\right) = -\frac{8\pi G}{c^4}T_{\mu\nu} \qquad (4.13)$$

$$\frac{1}{\left(\left|\Psi_o(r,t)\right|\right)^4} - 2\left(\frac{-6}{\left(\left|\Psi_i(r,t)\right|\right)^4}\right) = R - 2R = -\frac{8\pi G}{c^4}T_{\mu\nu} \qquad (4.14)$$

$$R - 2(R) = -R = -\frac{8\pi G}{c^4}T_{\mu\nu} \qquad (4.15)$$

$$-\frac{6}{\left(\left\|\Psi_i(r,t)\right\|\right)^4} = -\frac{8\pi G}{c^4}T_{\mu\nu} \tag{4.16}$$

$$6\left(\frac{\ddot{a}}{ac^2} + \frac{\dot{a}^2}{a^2c^2} + \frac{k}{a^2}\right) = \frac{8\pi G}{c^4}T_{\mu\nu} \tag{4.17}$$

$$G_{\mu\nu} = \frac{8\pi G}{c^4}T_{\mu\nu} \tag{4.18}$$

Expressing the wave function in six dimensions in terms of the Ricci curvature,

$$s = \frac{1}{\left(\left\|\Psi(r,t)\right\|\right)} = \frac{1}{\sqrt{r}} \cdot \frac{1}{\sqrt{t}} \tag{4.19}$$

$$s^4 = \frac{1}{\left(\left\|\Psi(r,t)\right\|\right)^4} = \frac{1}{r^2} \cdot \frac{1}{t^2} \tag{4.20}$$

Hence, the six-dimensional wavefunction EFEs emerge beautifully and natural to the complex geometry of space-time. The spatiotemporal variable "s" is inversely proportional to the curvature value of the wavefunction, and no longer directly proportional to a spatiotemporal surface "$r^2 \cdot t^2$" of a volume.

The inner and outer six-dimensional spatiotemporal cubes are given by

$$s_i^4 = \frac{r_i^2 \cdot t_i^2}{2} \tag{4.21}$$

$$s_o^4 = r_o^2 \cdot t_o^2 = \frac{1}{\left(\left\|\Psi_o(r,t)\right\|\right)^4} \tag{4.22}$$

The six-dimensional EFEs may be denoted as

$$R_{\mu\nu} - \frac{1}{5}g_{\mu\nu}R = \left(\frac{5}{5}\right)\frac{1}{\left(|\Psi_o(r,t)|\right)^4} - \frac{1}{5}(6)\left(\frac{-6}{\left(|\Psi_i(r,t)|\right)^4}\right) \qquad (4.23)$$

$$-\frac{R}{5} = -\frac{8\pi G}{c^5}T_{\mu\nu} \qquad (4.24)$$

$$\frac{6}{\left(|\Psi_i(r,t)|\right)^4} = \frac{20\pi G}{3c^5}T_{\mu\nu} \qquad (4.25)$$

$$6\left(\frac{\ddot{a}}{ac^2} + \frac{\dot{a}^2}{a^2c^2} + \frac{k}{a^2}\right) = \frac{20\pi G}{c^4}(-\rho+p) \qquad (4.26)$$

Therefore, the Einstein field equations are equations of the probability of the wave function of any object of mass or energy. The EFEs were conceptualized on static three-dimensional space-time and non-linear time was established as a folded three-dimensional temporal resultant (t^2) to fit the four-dimensional property of Minkowski space-time. Particles, objects of mass, or energy, distort space-time, and as the space-time gets distorted, the probability of spatiotemporal curvature emerges from the spatiotemporal wavelets, and the spatiotemporal curvature manifests the gravitational field about the geometry of the object or energy.

Probability, surface area, and curvature are intricately related in space-time. Quantum Mechanics was a part of General Relativity from the very beginning, or vice versa.

Firstly, let us consider the wave property of the outer spatiotemporal volume $r^2 \cdot t^2$ for the wave function,

$$\left(r^2 \cdot t^2\right)^i = s^i \qquad (4.27)$$

$$\left(s^2\right)^i = \left(e^{\ln s^{2i}}\right) = e^{i\ln s^2} \qquad (4.28)$$

$$\left(s^2\right)^i = \text{Cos}\left(\ln s^2\right) + i\,\text{Sin}\left(\ln s^2\right) \qquad (4.29)$$

$$\left(\frac{1}{\left(\left\|\Psi_o\left(r,t\right)\right\|\right)^4}\right)^i = \text{Cos}\left(\ln \frac{1}{\left(\left\|\Psi_o\left(r,t\right)\right\|\right)^4}\right) \qquad (4.30)$$

$$+i\,\text{Sin}\left(\ln \frac{1}{\left(\left\|\Psi_o\left(r,t\right)\right\|\right)^4}\right)$$

Secondly, let us consider the wave property of the inner spatiotemporal volume $\dfrac{r^2 \cdot t^2}{2}$ for the wave function,

$$\left(\frac{r^2 \cdot t^2}{2}\right)^i = \left(\frac{s^2}{2}\right)^i \qquad (4.31)$$

$$\left(\frac{s^2}{2}\right)^i = \left(e^{\ln\left(\frac{s^2}{2}\right)}\right)^i = e^{i\ln\left(\frac{s^2}{2}\right)} = e^{i\ln\left(\frac{1}{2\left(\left\|\Psi_o(r,t)\right\|\right)^4}\right)} \qquad (4.32)$$

$$\left(\frac{s^2}{2}\right)^i = \text{Cos}\left[\ln\left(\frac{s^2}{2}\right)\right] + i\,\text{Sin}\left[\ln\left(\frac{s^2}{2}\right)\right] \qquad (4.33)$$

$$\left(\frac{1}{2\left(\left\|\Psi_o\left(r,t\right)\right\|\right)^4}\right)^i = \text{Cos}\left[\ln \frac{1}{2\left(\left\|\Psi_o\left(r,t\right)\right\|\right)^4}\right] \qquad (4.34)$$

$$+i\,\text{Sin}\left[\ln \frac{1}{2\left(\left\|\Psi_o\left(r,t\right)\right\|\right)^4}\right]$$

Would integrating the inner spatiotemporal volume equation with respect to the wave function yield the outer spatiotemporal volume of the tesseract?

$$\int \left[\left(\frac{1}{2\left(\left\|\Psi_o\left(r,t\right)\right\|\right)^4} \right)^i d\left[2^{-1}\left(\left\|\Psi_o\left(r,t\right)\right\|\right)^{-4} \right] \right]^? = r^2 \cdot t^2 \qquad (4.35)$$

$$= \int \mathrm{Cos}\left[\ln \frac{1}{2\left(\left\|\Psi_o\left(r,t\right)\right\|\right)^4} \right] d\left[2^{-1}\left(\left\|\Psi_o\left(r,t\right)\right\|\right)^{-4} \right] \qquad (4.36)$$

$$+ i\int \mathrm{Sin}\left[\ln \frac{1}{2\left(\left\|\Psi_o\left(r,t\right)\right\|\right)^4} \right] d\left[2^{-1}\left(\left\|\Psi_o\left(r,t\right)\right\|\right)^{-4} \right]$$

$$= \left(\frac{1}{2\left(\left\|\Psi_o\left(r,t\right)\right\|\right)^4} \right)\left(\mathrm{Cos}\left[\ln \frac{1}{2\left(\left\|\Psi_o\left(r,t\right)\right\|\right)^4} \right] \qquad (4.37)$$

$$+ \mathrm{Sin}\left[\ln \frac{1}{2\left(\left\|\Psi_o\left(r,t\right)\right\|\right)^4} \right]$$

$$+ i\left(\mathrm{Cos}\left[\ln \frac{1}{2\left(\left\|\Psi_o\left(r,t\right)\right\|\right)^4} \right] - \mathrm{Sin}\left[\ln \frac{1}{2\left(\left\|\Psi_o\left(r,t\right)\right\|\right)^4} \right] \right)\right)$$

$$= \left(\frac{1}{2\left(\left\|\Psi_o\left(r,t\right)\right\|\right)^4} \right)\left(\mathrm{Cos}\left[\ln \frac{1}{2\left(\left\|\Psi_o\left(r,t\right)\right\|\right)^4} \right] \qquad (4.38)$$

$$+i\,\mathrm{Sin}\left[\ln\frac{1}{2\left(\left|\Psi_o\left(r,t\right)\right|\right)^4}\right]$$

$$-i\,\mathrm{Cos}\left[\ln\frac{1}{2\left(\left|\Psi_o\left(r,t\right)\right|\right)^4}\right]+\mathrm{Sin}\left[\ln\frac{1}{2\left(\left|\Psi_o\left(r,t\right)\right|\right)^4}\right]\right)$$

Simplifying terms, we get

$$\frac{1}{(1+i)}\left(\frac{1}{2\left(\left|\Psi_o\left(r,t\right)\right|\right)^4}\right)\left(\frac{1}{2\left(\left|\Psi_o\left(r,t\right)\right|\right)^4}\right)^i = \qquad (4.39)$$

$$=\left(\frac{1}{2\left(\left|\Psi_o\left(r,t\right)\right|\right)^4}\right)(1-i)\left(\mathrm{Cos}\left[\ln\frac{1}{2\left(\left|\Psi_o\left(r,t\right)\right|\right)^4}\right]+i\,\mathrm{Sin}\left[\ln\frac{1}{2\left(\left|\Psi_o\left(r,t\right)\right|\right)^4}\right]\right)$$

$$\frac{1}{(1+i)}\int\left(\frac{1}{2\left(\left|\Psi_o\left(r,t\right)\right|\right)^4}\right)^{i+1}d\left(2^{-1}\left(\left|\Psi_o\left(r,t\right)\right|\right)^{-4}\right)\qquad (4.40)$$

$$=\frac{1}{(1+i)}\left(\frac{1}{2\left(\left|\Psi_o\left(r,t\right)\right|\right)^4}\right)^{i+1}+C$$

$$=\int\mathrm{Cos}\left[\ln\frac{1}{2\left(\left|\Psi_o\left(r,t\right)\right|\right)^4}\right]d\left[2^{-1}\left(\left|\Psi_o\left(r,t\right)\right|\right)^{-4}\right]$$

$$+i\int\mathrm{Sin}\left[\ln\frac{1}{2\left(\left|\Psi_o\left(r,t\right)\right|\right)^4}\right]d\left[2^{-1}\left(\left|\Psi_o\left(r,t\right)\right|\right)^{-4}\right]$$

$$= \text{Cos}\left[\ln\frac{1}{2\left(\left\|\Psi_o\left(r,t\right)\right\|\right)^4}\right] + i\,\text{Sin}\left[\ln\frac{1}{2\left(\left\|\Psi_o\left(r,t\right)\right\|\right)^4}\right]$$

$$= \left(\frac{1}{2\left(\left\|\Psi_o\left(r,t\right)\right\|\right)^4}\right)^i = r^2\cdot t^2$$

The proof of the pudding is in the waving of the function!

Therefore, the spatiotemporal wavefunction of six-dimensional space-time may be interpreted as the underlying infrastructure to all there is in our universe, space, time, mass, and all forms of energy. The wave function is probability, wave property, proportionality, curvature, gravitation, complex space-time, in the concept of Special and General Relativity, part of the EFEs, part of the Schrödinger's equation, ubiquitous at any physical scale, part of the Pauli exclusion principle, and part of the exponent of the kernel of growth of all natural things. The wavefunction is the centerpiece of modern physics!

§ 5. What is the principle of equivalence between pressure and energy density? How is the wave function related to the pressure of spatiotemporal divergence?

The state of a quantum object is completely specified by a complex wave function $\Psi(x)$, which is a single-valued function of position. A single-valued wave function is a function that, for each value of "x" in its domain, has a unique value in its range. The probability density that the object will be found at position "x" is determined by $\left|\Psi(x)\right|^2 = \left|-1/\ln\ e^{-x}\right|^2 = 1/x$. Spatial length "$x$" is equal to the reciprocal of the squared wave function of an object, $x = 1/\left[\Psi(x)\right]^2$.

The divergence of the spatiotemporal wave function is the extent to which the wave function vector field flux behaves like a source at a given point.

The three-dimensional wave function vector is given by

$$\vec{\Psi} = \Psi_x \vec{e}_x + \Psi_y \vec{e}_y + \Psi_z \vec{e}_z \tag{5.1}$$

and the six-dimensional wave function vector is denoted as

$$\vec{\Psi}_{st} = -\Psi_{t_x}\vec{e}_{t_x} - \Psi_{t_y}\vec{e}_{t_y} - \Psi_{t_z}\vec{e}_{t_z} + \Psi_x\vec{e}_x + \Psi_y\vec{e}_y + \Psi_z\vec{e}_z \tag{5.2}$$

In three-dimensional Cartesian coordinates, the divergence of the continuously differentiable wave function vector field is defined as a function with a scalar value.

$$div\ \vec{\Psi} = \nabla \cdot \vec{\Psi} = \frac{\partial \Psi_x}{\partial x} + \frac{\partial \Psi_y}{\partial y} + \frac{\partial \Psi_z}{\partial z} \tag{5.3}$$

From previous research, a characteristic of the spatiotemporal wave function vector field is that its n-divergence outside its physical source for an irrotational vector field in six-dimensional space-time is not zero, $div^n\ \vec{\Psi}_{st} = \Re \cdot \vec{\Psi}_{st} \neq 0$. (Nieves, 2020)

$$div^n\ \vec{\Psi}_{st} = \Re \cdot \vec{\Psi}_{st} = \frac{1}{c}\frac{\partial \Psi_{t_x}}{\partial t_x} + \frac{1}{c}\frac{\partial \Psi_{t_y}}{\partial t_y} + \frac{1}{c}\frac{\partial \Psi_{t_z}}{\partial t_z} \tag{5.4}$$

$$+\frac{\partial \Psi_x}{\partial x} + \frac{\partial \Psi_y}{\partial y} + \frac{\partial \Psi_z}{\partial z}$$

Hence, every spatiotemporal point has a scalar value for its spatiotemporal divergence.

The following principle of equivalence between pressure and energy density is the theoretical framework of the Einstein Field Equations for the General Theory of Relativity:

$$\text{Pressure} \equiv \text{Energy Density} \tag{5.5}$$

The following equation is a time independent pressure-to-energy density equation for the "Quantum Mechanical − General Relativity" attribute of the wave function at a point $P(x, y, z)$:

$$\nabla^2 \Psi = \frac{Gc^2 \rho}{c^4} \Psi = \frac{G\rho}{c^2} \Psi \tag{5.6}$$

$$\frac{\partial^2 \Psi}{\partial x^2} + \frac{\partial^2 \Psi}{\partial y^2} + \frac{\partial^2 \Psi}{\partial z^2} = \frac{c^3 \rho}{h\omega^2} \Psi \tag{5.7}$$

Where Ψ is the wave function defined over space, ∇^2 is the Laplacian operator for three-dimensional space, $c^2\rho$ is the energy density of mass, g is the gravitational acceleration, G is Newton's constant, c is the speed of light, ω is the angular frequency, and h is the Planck constant.

The pressure-to-energy density equation defined over space and time, for four dimensions is given by

$$\Box\Psi(r,t) = \frac{G\rho}{c^2} \Psi(r,t) \tag{5.8}$$

or for six dimensions, it would be

$$\vec{\Re}^2 \Psi_{st}(r,t) = \frac{G\rho}{c^2} \Psi_{st}(r,t) \tag{5.9}$$

Where the symbol $\Box$ is the four-dimensional d'Alembert operator, or wave operator, and $\vec{\Re}^2$, or $\Diamond^2$, is the six-dimensional double, or square, Robertonian operator, and ρ is the mass density.

Is there a dynamic field format to represent the constants in the Einstein EFEs?

Let us discuss the present format of the Einstein field equations for General Relativity in the absence of cosmological energy.

$$R_{\mu\nu} - \frac{1}{(n-1)} g_{\mu\nu} R = \frac{8\pi G}{c^4} (T_{\mu\nu}) \tag{5.10}$$

The constants "*G*" and "*c*" are used in the present format to represent a static force in the Einstein constant. Even though these constants are relativistic, they are invariant to position in the coordinate system of the gravitational field of the mass. Hence, the values over time of quantities of mass, volume, space, and time, in the constants, are not adjusted to the specific point in the space-time-mass of the system. However useful this may sound, for instance, a GPS satellite system would need to have a built-in, or integral system, that measures those quantities and calculates their location in order to be highly accurate. The mass is a constantly varying quantity, the jolt of the volume is transcendental, length is an average radial distance, and time is presently a calculated quantity between two clocks that is at best a relativistic approximation of a temporal distance and a radial curvature. All these qualities of the physical system and calculations add to the uncertainty and inaccuracy of the GPS system.

It is proposed that by using the field potential of the gravitational field of the mass, and the jolt of the volume, an integral part of the GPS system would be measuring the field potential, and the jolt of the volume, at the point of measurement, and a faster calculation may be made that is highly accurate and self-reliant for each satellite in the system. The entire GPS system will be more reliant and accurate since each satellite will be more independently reliable, accurate, cost effective, and hardened, to reduce the probability of a system wide failure.

Let us propose the following format for the Einstein EFEs,

$$R_{\mu\nu} - \frac{1}{(n-1)} g_{\mu\nu} R = \frac{8\pi}{\nabla V(r,t)} \left(T_{\mu\nu} \right) \tag{5.11}$$

$$G_{\mu\nu} = \frac{8\pi}{\nabla V(r,t)} T_{\mu\nu} \tag{5.12}$$

$$\nabla V(r,t) G_{\mu\nu} = 8\pi T_{\mu\nu} \tag{5.13}$$

$$\nabla V(r,t) G_{\mu\nu} = \nabla^3 a \left(T_{\mu\nu} \right) \tag{5.14}$$

Where $V(r,t)$ is the field potential, "a" is the volume of the mass, and $\nabla^3 a$ is the jolt of the volume.

Therefore, the Einstein EFEs are fully dynamic equations in this format. Let us name the variable (Gimel) "ג" $\equiv \nabla^3 a / \nabla V(r,t)$, the Grossmann variable, in honor of Marcel Grossmann, an eminent mathematician, a classmate, and a crucial collaborator of Einstein on the General Theory of Relativity.

PART III

QUANTUM MECHANICS

Chapter 4

Time and Space are Nonlinear

§ 1. How does the human mind measure time? Is the human mind using more than one clock?

Let us imagine an alternative universe of solipsism where our concept of time is a specific temporal coordinate point in which our consciousness travels through the vastness of space-time within the dynamic block universe where all other entities are going about their actions with our consciousness being the only one that is actually there because all the other consciousness are located at a different temporal coordinate point so that every consciousness is actually the only protagonist of the stage of its reality at a given instant in time. All other consciousness have already acted their part of their history while from the point of view of every single consciousness it is all now just happening in real time while all other consciousness from their perspective have already been there and done their part in their unique instant of time. Will this type of reality be similar to the experience that an individual consciousness may experience in a simulated holographic reality if that were possible? Would the measurement of time be the same in such a universe from the perspective of an individual mind? Would the measurement of time for an individual mind be provided externally?

§ 2. The Light Cone.

A light cone in the Special and General Theory of Relativity, is the path that a beam of light from an event at a spatiotemporal point, $P(x, y, z)$, would follow at the speed of light traveling in all possible spatiotemporal directions. It is possible to visualize the light cone as a photon that expands towards the future or the past at the speed of light.

In Minkowski space-time or in a hyperbolic universe with spatiotemporal curvature, the past light cone is the boundary of the causal past and the future light cone is the boundary of the causal future. Nevertheless, in a region of space-time where there is gravitational lensing, the light cone may double over itself, to such a degree that part of the cone may be within its causal past or its causal future, and no longer on the boundary. Hence, natural spatiotemporal curvature can bend light cones and cause spatiotemporal loops to its future or past. (Rucker, 1977)

In curved space-time, the light cones cannot all be parallel to each other like in flat space-time, so they cannot all be equally tilted. The non-vanishing Weyl tensor exhibits the characteristic of not allowing all light cones to be parallel in free space (no matter). Since, spatiotemporal curvature can distort the shape of a light cone, so that its boundary is no longer at 45 degrees, as in the case of a curved elliptical light cone, let us consider a circular cone in a nearly flat spatiotemporal region.

Furthermore, let us visualize a six-dimensional light cone where all the spatial dimensions are expanding or contracting, as each temporal dimension expands creating more space. Hence, the 45 degree boundary lines are shifting through spatial distances as the light cone gradually widens in all its dimensions, and the point of the observer extends or contracts, as a spatiotemporal sphere of an expanded or contracted light cone. The future light cone would be gradually larger than its past cone in an expanding universe.
For a circular cone in six-dimensional nearly flat space-time, we have

$$ds^2 = dx_\mu^2 + dx_\nu^2 + dx_\sigma^2 - c^2 dt_\mu^2 - c^2 dt_\nu^2 - c^2 dt_\sigma^2 \qquad (2.1)$$

For the null geodesics with $ds^2 = 0$, $dx_\sigma^2 = 0$, and with time folded, this equation reduces to the familiar equation of a circle.

$$0 = ds^2 = dx_\mu^2 + dx_\nu^2 - c^2 dt_\phi^2 \qquad (2.2)$$

$$c^2 dt_\phi^2 = dx_\mu^2 + dx_\nu^2 \qquad (2.3)$$

Let us consider a curved space-time for a relativistic light cone, so through parallel transport the curvature of the spatiotemporal light

66

cone may be described through the Christoffel symbols and the Riemann curvature tensors for space and time. The spatiotemporal assumption is that for each spatial dimension there is a conjugate temporal dimension that may also be curved in six-dimensional space-time.

The Christoffel symbols are also known as the affine connection coefficients or the Levi-Civita connection coefficients; they are symmetric in the two lower indices.

The Spatial Christoffel Symbol

$$\Gamma^{\lambda}_{\mu\nu} = \frac{g^{\lambda\sigma}}{2}\left(\frac{dg_{\sigma\mu}}{dx^{\nu}} + \frac{dg_{\sigma\nu}}{dx^{\mu}} - \frac{dg_{\mu\nu}}{dx^{\sigma}}\right) \tag{2.4}$$

The Temporal Christoffel Symbol

$$\Pi^{\lambda}_{\mu\nu} = -\frac{g^{\lambda\sigma}}{2}\left(\frac{dg_{\sigma\mu}}{dt^{\nu}} + \frac{dg_{\sigma\nu}}{dt^{\mu}} - \frac{dg_{\mu\nu}}{dt^{\sigma}}\right) \tag{2.5}$$

$$= \frac{h^{\lambda\sigma}}{2}\left(\frac{dh_{\sigma\mu}}{dt^{\nu}} + \frac{dh_{\sigma\nu}}{dt^{\mu}} - \frac{dh_{\mu\nu}}{dt^{\sigma}}\right)$$

The Riemann Spatial Curvature Tensor

$$R^{\lambda}_{\sigma\mu\nu} = \frac{d\Gamma^{\lambda}_{\sigma\nu}}{dx^{\mu}} + \Gamma^{\alpha}_{\sigma\nu}\Gamma^{\lambda}_{\alpha\mu} - \frac{d\Gamma^{\lambda}_{\sigma\mu}}{dx^{\nu}} - \Gamma^{\alpha}_{\sigma\mu}\Gamma^{\lambda}_{\alpha\nu} \tag{2.6}$$

The Riemann Temporal Curvature Tensor

$$H^{\lambda}_{\sigma\mu\nu} = -\frac{d\Pi^{\lambda}_{\sigma\nu}}{dx^{\mu}} - \Pi^{\alpha}_{\sigma\nu}\Pi^{\lambda}_{\alpha\mu} + \frac{d\Pi^{\lambda}_{\sigma\mu}}{dx^{\nu}} + \Pi^{\alpha}_{\sigma\mu}\Pi^{\lambda}_{\alpha\nu} \tag{2.7}$$

Therefore, the spatiotemporal curvature equation for the six-dimensional light cone may be expressed as

$$\Psi^{\lambda}_{\sigma\mu\nu} = R^{\lambda}_{\sigma\mu\nu} + H^{\lambda}_{\sigma\mu\nu} \tag{2.8}$$

$$\Psi^{\lambda}{}_{\sigma\mu\nu} = \frac{d\Gamma^{\lambda}{}_{\sigma\nu}}{dx^{\mu}} + \Gamma^{\alpha}{}_{\sigma\nu}\Gamma^{\lambda}{}_{\alpha\mu} - \frac{d\Gamma^{\lambda}{}_{\sigma\mu}}{dx^{\nu}} - \Gamma^{\alpha}{}_{\sigma\mu}\Gamma^{\lambda}{}_{\alpha\nu} \qquad (2.9)$$

$$- \frac{d\Pi^{\lambda}{}_{\sigma\nu}}{dx^{\mu}} - \Pi^{\alpha}{}_{\sigma\nu}\Pi^{\lambda}{}_{\alpha\mu} + \frac{d\Pi^{\lambda}{}_{\sigma\mu}}{dx^{\nu}} + \Pi^{\alpha}{}_{\sigma\mu}\Pi^{\lambda}{}_{\alpha\nu}$$

$$\Psi^{\lambda}{}_{\sigma\mu\nu} = \frac{d\Gamma^{\lambda}{}_{\sigma\nu}}{dx^{\mu}} + \Gamma^{\alpha}{}_{\sigma\nu}\Gamma^{\lambda}{}_{\alpha\mu} + \frac{d\Pi^{\lambda}{}_{\sigma\mu}}{dx^{\nu}} + \Pi^{\alpha}{}_{\sigma\mu}\Pi^{\lambda}{}_{\alpha\nu} \qquad (2.10)$$

$$- \frac{d\Gamma^{\lambda}{}_{\sigma\mu}}{dx^{\nu}} - \Gamma^{\alpha}{}_{\sigma\mu}\Gamma^{\lambda}{}_{\alpha\nu} - \frac{d\Pi^{\lambda}{}_{\sigma\nu}}{dx^{\mu}} - \Pi^{\alpha}{}_{\sigma\nu}\Pi^{\lambda}{}_{\alpha\mu}$$

It is possible to think of other light cones as the light cones of other worldlines that may be coincident, non-coincident, parallel, tilted, curved, or a combination of these states. The space beyond the 45 degree lines of a vertical light cone may be occupied by other light cones that are perpendicular whose causal past or future may be unrelated to the vertical light cone. Those perpendicular light cones may also have the aforementioned states which may change the causality between light cones.

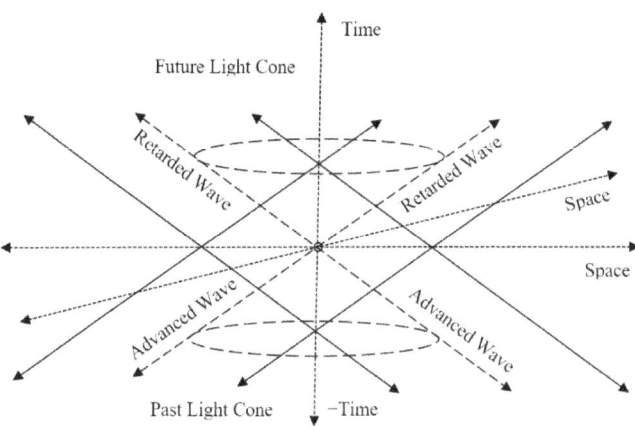

Figure 1. The spatiotemporal light cone

The conic shape of a light cone exemplifies the spatiotemporal expansion or contraction. The future light cone is all the possible forward wavefunctions, and the past light cone is all the possible backward wavefunctions. The world line of the observer is the observer's wavefunction. At the center point, or at any other point, of the light cone, there are two six-dimensional waves present simultaneously, the forward wavefunction and the backward wavefunction, depending on the phase of the wave of the observer or the object, the observer or object selects what wave would carry the observer forward or backward. When an object travels on the temporal wave, its clock does not tick tock, but space seems to be flowing through the object at *"c"*, what would make the body of the object change shape or volume depending on what space is doing, that is, expanding, static, or contracting. On the other hand, when the object travels on the spatial wave, time flows through the object at *"c"*, what would make the object's clock tick tock faster, slower, or not at all, depending on what time is doing, that is, contracting, dilating, or static.

The center point of a light cone is the location of an observer. The observer may travel in any of the six directions of space-time. Each spatial dimension has two opposite directions, and each dimension has an orthogonal conjugate temporal dimension in six-dimensional space-time. In this manner, every spatial direction has a conjugate temporal direction. When an observer makes a right turn, the future light cone is tilted 90 degrees, turned on its side, to a causal future elsewhere in space, perpendicular to the original trajectory of the worldline of the observer. The light cones of parallel worlds in the many worlds interpretation may be located in a different space-time that has its own frequency, wavelength, phase, orientational states, and spatiotemporal location in the bulk. Parallel worlds that are similar are more coincidental in their nature. There may be spatiotemporal bridges across parallel worlds or within the same world as predicted by the General Theory of Relativity.

§ 3. Natural spatiotemporal waves.

It is interesting to note that when natural numbers are plotted on a four-dimensional surface, with coordinates (x, y, t_X, t_Y) and polar coordinates *(r, θ)*, represent a counterclockwise Archimedean spiral when equally spaced radial and radian distances are graphed or projected on a homogenous and isotropic spatiotemporal coordinate grid. Natural numbers illustrate the expansion of space-time as natural spatiotemporal waves expand radially outward at equal temporal radian distances. Prime numbers also represent two clockwise prime spatiotemporal spirals, one prime spatiotemporal spiral that expands upward and another prime spatiotemporal spiral that expands downward. A prime spatiotemporal wave is a natural wave greater than a unitary spatiotemporal wave that cannot be formed by multiplying two smaller natural spatiotemporal waves. A natural spatiotemporal wave greater than a unitary spatiotemporal wave that is not a prime spatiotemporal wave is a composite spatiotemporal wave.

At a short distance above the grid, there are two bundles of ten prime spatiotemporal spirals. As the distance increases, seventy bundles of four prime spatiotemporal spirals, for a total of two hundred and eighty prime spatiotemporal spirals, are radially discernable. It is interesting to note that two hundred and eighty divided by twenty is very close to 2π. This graphical representation illustrates that the transcendental number π is fundamentally related to the spatiotemporal expansion.

A natural spatiotemporal spiral, that has all multiples of six *(6x)*, represents the six spatiotemporal dimensions and directions, while any other natural spatiotemporal spiral is one unitary spatiotemporal wave above the previous one. In that respect, the 6th natural spatiotemporal spiral is fundamental and transcendental in nature. Every full counterclockwise rotation from the fundamental falls on the same fundamental spatiotemporal wave as it continues to expand as *(6x + n)* where "*n*" is a natural number or zero. Thus, all prime spatiotemporal waves are a multiple spatiotemporal wave, or five unitary spatiotemporal waves, above the fundamental.

Aside, Dirichlet's Theorem predicts that one fourth of each of the proportions of prime numbers would end in 1, 3, 7, or 9.

§ 4. *Does the geodesic equation of motion follow from the Einstein field equations for empty space?*

The original EFEs were incomplete from the start as a field theory because they relied on the independent postulate that the geodesic equation of motion was the law of motion for a particle, and not the

EFEs themselves. (Einstein, 1935) However, the eminent physicist Albert Einstein believed that the geodesic equation of motion could be derived from his field equations for empty space since the Ricci curvature tensor vanishes in empty space. (Einstein, 2003) A complete field theory encompasses fields, particles, and motion, as parts of the theory.

Proper space or proper time are relativistic and nonlinear. Objects of mass or energy do not indubitably travel in a linear trajectory through space. Objects of mass or energy are inclined to travel on a curved trajectory, as a wave or cycle, such as a helical trajectory, or randomly, and as determined by the laws of motion of our universe described by the geodesic equation of motion. Observation of a moving object of mass or energy may qualify itself as a spatiotemporal measurement. There may be a change of phase and/or momentum between the observer and what is observed. Consequently, an observer's point of view on an object determines the proper spatial distance between the observer and the moving object being observed. Proper time may be plainly described as an observer's random measurement of the motion of objects of mass or energy through proper space.

Space-time has a uniform cycle for any wave function. The uniform cycle of a wave function is always 2π while the proper space and the proper time of a wave function are relativistic. The cycle of the wave function is absolute time in radians, or absolute temporal units, and may be considered Newton's clock for absolute time. This universal clock encompasses all relativistic clocks within its temporal realm. The universal clock is the fundamental wave function equal to the sum of all relativistic wave function harmonics. All potential relativistic frequencies emerge eventfully from the fundamental. The integral sum of the discrete relativistic frequencies that manifests each dimension in the frequency domain of reality materialize at any observable scale the existence of space-time.

Let us imagine the clock dial in one of the most famous clock towers in the world in a thought experiment, the clock tower at the Palace of Westminster in London, U.K., which houses the famous great bell of the striking clock at the north end known as Big Ben. The great bell is so famous that the name is frequently extended to refer to both the clock and the clock tower. The beautiful clock dial was designed by

Augustus Pugin and completed in 1859. Each clock dial is 7 meters (23 feet) in diameter. A moving mechanical clock measures proper time.

Let the Big Ben clock dial represent an imaginary stationary clock in space-time that is not in a gravitational field, in isotropic and homogeneous space-time, with a particle emitter at its center so that it can emit a particle that travels in a helical fashion from the dial in a way equal to the cyclical movement of the hands of the clock, so that we can observe the trajectory of the wave of the particle as if the diameter of the wave were the same diameter as the clock dial with negligible expansion of the helical trajectory. Thus, the amplitude of the wave represents a proper length, the distance of propagation away from the dial represents proper time, the cyclical movement in radians of the hands of Big Ben represent absolute time, and the ratio of the circumference to the diameter of the dial is absolute flat space. This relationship is also found in the geometry of the spatiotemporal wave between the absolute circumference and the relativistic spatial and temporal distances of the displacement. Absolute space and absolute time are transcendental.

If the dial were moving in the strong gravitational field of a planet, the amplitude of the wave of the emitted particle would increase, the wavelength would decrease, time would dilate, and the frequency would increase, but the cyclical movement in radians of the hands of Big Ben, or helical movement of the wave, would remain the same, that is, 2π/wavelength, regardless of scale. Since the dial changes as the amplitude, then the dial would seem extended. The amplitude extends in the temporal dimension that is orthogonal to the spatial dimension in the direction of propagation. Every spatial dimension has a conjugate temporal dimension.

If the dial were moving outside the gravitational field of a planet, the amplitude of the wave of the emitted particle would decrease, the wavelength would increase, time would pass faster, and the frequency would decrease, but the cyclical movement in radians of the hands of Big Ben, or helical movement of the wave, would remain the same. Since the dial changes as the amplitude, then the dial would seem contracted.

Hence, absolute time may be viewed as a fundamental oscillation or frequency that is uniform throughout our entire universe regardless of scale.

A geodesic gives a general form to the idea of a straight line to the trajectory of an object or particle through curved space-time according to the General Theory of Relativity. The trajectory is spatial and temporal in its nature, as space and time are reciprocal.

The spatiotemporal worldline of an object or particle, in the absence of any non-gravitational forces, is a specific kind of geodesic; it is a spatiotemporal geodesic. Any falling or freely moving object or particle would always follow the path of a geodesic. (Einstein, 1952)

The proper spatial distance for a moving object or particle may be expressed as

$$r = \sqrt[2]{x^2 + y^2 + z^2} \qquad (4.1)$$

The proper temporal distance for a moving object or particle may be expressed as

$$\tau = \sqrt[2]{\left(\tau_x\right)^2 + \left(\tau_y\right)^2 + \left(\tau_z\right)^2} \qquad (4.2)$$

The spatiotemporal geodesic equation is the path of the observer through his worldline.

The spatial geodesic equation with folded proper space is

$$\frac{d^2 x^\lambda}{dr^2} + \Gamma^\lambda_{\mu\nu} \frac{dx^\mu}{dr} \frac{dx^\nu}{dr} = 0 \qquad (4.3)$$

The spatial geodesic equation with unfolded proper space is

$$\frac{d^2 x^\lambda}{dx^2} + \frac{d^2 x^\lambda}{dy^2} + \frac{d^2 x^\lambda}{dz^2} \qquad (4.4)$$

73

$$+\left(\Gamma^{\lambda}_{\ \mu\nu}\right)\left(\frac{dx^{\mu}}{dx}+\frac{dx^{\mu}}{dy}+\frac{dx^{\mu}}{dz}\right)\left(\frac{dx^{\nu}}{dx}+\frac{dx^{\nu}}{dy}+\frac{dx^{\nu}}{dz}\right)=0$$

The temporal geodesic equation with folded proper time is

$$-\frac{d^2x^{\lambda}}{d\tau^2}-\Gamma^{\lambda}_{\ \mu\nu}\frac{dx^{\mu}}{d\tau}\frac{dx^{\nu}}{d\tau}=0 \qquad (4.5)$$

The temporal geodesic equation with unfolded proper time is

$$-\frac{d^2x^{\lambda}}{d\tau_x^2}-\frac{d^2x^{\lambda}}{d\tau_y^2}-\frac{d^2x^{\lambda}}{d\tau_z^2} \qquad (4.6)$$

$$-\Gamma^{\lambda}_{\ \mu\nu}\left(\frac{dx^{\mu}}{d\tau_x}+\frac{dx^{\mu}}{d\tau_y}+\frac{dx^{\mu}}{d\tau_z}\right)\left(\frac{dx^{\nu}}{d\tau_x}+\frac{dx^{\nu}}{d\tau_y}+\frac{dx^{\nu}}{d\tau_z}\right)=0$$

Therefore, the spatiotemporal geodesic equation of motion with folded proper space and folded proper time may be expressed as

$$\frac{d^2x^{\lambda}}{dr^2}-\frac{1}{c^2}\frac{d^2x^{\lambda}}{d\tau^2}+\Gamma^{\lambda}_{\ \mu\nu}\left(\frac{dx^{\mu}}{dr}\frac{dx^{\nu}}{dr}-\frac{1}{c^2}\frac{dx^{\mu}}{d\tau}\frac{dx^{\nu}}{d\tau}\right)=0 \qquad (4.7)$$

Rearranging the spatiotemporal geodesic equation into spatial or temporal terms, we have

$$\left(\frac{d^2x^{\lambda}}{dr^2}+\Gamma^{\lambda}_{\ \mu\nu}\frac{dx^{\mu}}{dr}\frac{dx^{\nu}}{dr}\right) \qquad (4.8)$$

$$+\left(-\frac{1}{c^2}\frac{d^2x^{\lambda}}{d\tau^2}-\Gamma^{\lambda}_{\ \mu\nu}\frac{1}{c^2}\frac{dx^{\mu}}{d\tau}\frac{dx^{\nu}}{d\tau}\right)=0$$

Hence, let us describe the Christoffel symbol for the metric tensor in Minkowski space-time where the spatial curvature and the temporal curvature equal zero, or in the concrete case that the spatiotemporal curvature is nearly zero, in empty space.

For flat space,

$$R_{\mu\nu} - \frac{1}{(n-1)} g_{\mu\nu} R \equiv 0 \qquad (4.9)$$

For flat time,

$$-H_{\mu\nu} - \frac{1}{(n-1)} h_{\mu\nu} H \equiv 0 \qquad (4.10)$$

The spatial Christoffel symbol is

$$\Gamma^\lambda{}_{\mu\nu} = \frac{g^{\lambda\sigma}}{2} \left(\frac{dg_{\sigma\mu}}{dx^\nu} + \frac{dg_{\sigma\nu}}{dx^\mu} - \frac{dg_{\mu\nu}}{dx^\sigma} \right) \qquad (4.11)$$

$$\Gamma^\lambda{}_{\mu\nu} = \frac{g^{\lambda\sigma}}{2} \left(\frac{dg_{\sigma\mu}}{dx^\nu} + \frac{dg_{\sigma\nu}}{dx^\mu} - \frac{dg_{\mu\nu}}{dx^\sigma} \right) \qquad (4.12)$$

$$= \frac{1}{(n-1)} g^\lambda{}_{\mu\nu} g^{\sigma\mu\nu} \left(\frac{dg_{\sigma\mu}}{dx^\nu} + \frac{dg_{\sigma\nu}}{dx^\mu} - \frac{dg_{\mu\nu}}{dx^\sigma} \right)$$

$$\Gamma^\lambda{}_{\mu\nu} = \frac{1}{(n-1)} \left(g^\lambda{}_{\mu\nu} \cdot \vec{e}_\lambda \right) \left(\frac{dg^\nu}{dx^\nu} + \frac{dg^\mu}{dx^\mu} - \frac{dg^\sigma}{dx^\sigma} \right) \qquad (4.13)$$

$$= \frac{1}{(n-1)} \left(g_{\mu\nu} \cdot \vec{e} \right) \left(\vec{e} \right) = \frac{1}{(n-1)} \left(g_{\mu\nu} \right) \left(\vec{e} \cdot \vec{e} \right)$$

Where $\vec{e}_\lambda$ is the vector of parallel transport.

$$\Gamma^\lambda{}_{\mu\nu} = \frac{1}{(n-1)} \left(g_{\mu\nu} \right) (1) = \frac{1}{(n-1)} \left(g_{\mu\nu} \right) \qquad (4.14)$$

The temporal Christoffel symbol is

$$\Pi^{\lambda}{}_{\mu\nu} = -\Gamma^{\lambda}{}_{\mu\nu} = \frac{h^{\lambda\sigma}}{2}\left(\frac{dh_{\sigma\mu}}{dx^{\nu}} + \frac{dh_{\sigma\nu}}{dx^{\mu}} - \frac{dh_{\mu\nu}}{dx^{\sigma}}\right) \quad (4.15)$$

$$= \frac{1}{(n-1)}h^{\lambda}{}_{\mu\nu}h^{\sigma\mu\nu}\left(\frac{dh_{\sigma\mu}}{dx^{\nu}} + \frac{dh_{\sigma\nu}}{dx^{\mu}} - \frac{dh_{\mu\nu}}{dx^{\sigma}}\right)$$

$$\Pi^{\lambda}{}_{\mu\nu} = -\Gamma^{\lambda}{}_{\mu\nu} = \frac{h^{\lambda\sigma}}{2}\left(\frac{dh_{\sigma\mu}}{dx^{\nu}} + \frac{dh_{\sigma\nu}}{dx^{\mu}} - \frac{dh_{\mu\nu}}{dx^{\sigma}}\right) \quad (4.16)$$

$$= \frac{1}{(n-1)}h^{\lambda}{}_{\mu\nu}h^{\sigma\mu\nu}\left(\frac{1}{c^2}\right)\left(\frac{dh_{\sigma\mu}}{d\tau^{\nu}} + \frac{dh_{\sigma\nu}}{d\tau^{\mu}} - \frac{dh_{\mu\nu}}{d\tau^{\sigma}}\right)$$

Following a similar derivation for the temporal Christoffel symbol, we obtain

$$\Pi^{\lambda}{}_{\mu\nu} = -\Gamma^{\lambda}{}_{\mu\nu} = \frac{1}{(n-1)}\left(h_{\mu\nu}\right)\left(\frac{1}{c^2}\right)(1) = \frac{1}{(n-1)}\left(h_{\mu\nu}\right)\left(\frac{1}{c^2}\right) \quad (4.17)$$

The spatial metric tensor "$g_{\mu\nu}$" was adjusted to be four-dimensional even though the Ricci curvature was formulated for three-dimensional space. The temporal metric tensor "$h_{\mu\nu}$" was adjusted to be three-dimensional as a bona-fide fudge factor since the temporal dimension was considered, to be a single dimension in Minkowski space-time or just a magnitude, as it is at the time of this writing.

It is worthy to point out that it was a prodigious effort of Albert Einstein, even with the help and teaching of his classmate, the eminent mathematician Marcel Grossmann, on differential geometry, to be the first to formulate the original EFEs when the understanding of space-time and other modern concepts of science were in their very early stages. Moreover, Einstein was under a lot of pressure to formulate his EFEs before David Hilbert who was hard on his heels. Hilbert was one of the most influential, brilliant, and universal mathematicians of the 19th and early 20th centuries. As a matter of fact, Hilbert arrived at the EFEs through an axiomatic derivation of the field equations, the Einstein–Hilbert action.

In the original EFEs, $n = 3$ for spatial curvature and $n = 4$ for temporal curvature. Space was considered to be static, and not expanding, and time was the magnitude of the fourth dimension. Thus, three equal components were added as time-time components to Minkowski space-time.

$$\left(\frac{d^2x^\lambda}{dr^2} + \frac{1}{(n-1)}(g_{\mu\nu})\left[\frac{dx^\mu}{dr}\frac{dx^\nu}{dr} \right] \right)$$

(4.18)

$$+3\left(-\frac{1}{c^2}\frac{d^2x^\lambda}{d\tau^2} - \frac{1}{(n-1)}(h_{\mu\nu})\frac{1}{c^2}\left[\frac{dx^\mu}{d\tau}\frac{dx^\nu}{d\tau} \right] \right) = 0$$

$$\left(\frac{d^2x^\lambda}{dr^2} + \frac{1}{2}g_{\mu\nu}\frac{dx^\mu}{dr}\frac{dx^\nu}{dr} \right)$$

(4.19)

$$+3\left(-\frac{1}{c^2}\frac{d^2x^\lambda}{d\tau^2} - \frac{1}{3}h_{\mu\nu}\frac{1}{c^2}\frac{dx^\mu}{d\tau}\frac{dx^\nu}{d\tau} \right) = 0$$

Changing spatiotemporal coordinate and tensor notation to algebraic symbols, where "Σ" is coordinate space, "σ" is proper space, "τ" is proper time, "g" is the trace of the spatial metric tensor, "h" is the trace of the temporal metric tensor, "a" is a volume of space, "c" is the speed of light, and "k" is a constant of spatial or temporal curvature, which in the case of empty space they may be the same.

$$\left(\frac{d^2\Sigma}{d\sigma^2} + \frac{1}{2}(g)\left(\frac{d\Sigma}{d\sigma} \right)^2 + k_s \right)$$

(4.20)

$$-3\left(\frac{1}{c^2}\frac{d^2\Sigma}{d\tau^2} + \frac{1}{3}(h)\left(\frac{1}{c}\frac{d\Sigma}{d\tau} \right)^2 + ck_\tau \right) = 0$$

$$\left(\frac{d^2\Sigma}{d\sigma^2} + \frac{1}{2}(4)\left(\frac{d\Sigma}{d\sigma}\right)^2 + k\right) \qquad (4.21)$$

$$-3\left(\frac{1}{c^2}\frac{d^2\Sigma}{d\tau^2} + \frac{1}{3}(3)\left(\frac{1}{c}\frac{d\Sigma}{d\tau}\right)^2 + k\right) = 0$$

$$\frac{\dot{a}^2}{c^2} + \frac{2\ddot{a}a}{c^2} + k - \frac{3\dot{a}^2}{c^2} - \frac{3\ddot{a}a}{c^2} - 3k = 0 \qquad (4.22)$$

$$\frac{\dot{a}^2}{a^2c^2} + \frac{2\ddot{a}}{ac^2} + \frac{k}{a^2} - \frac{3\dot{a}^2}{a^2c^2} - \frac{3\ddot{a}}{ac^2} - \frac{3k}{a^2} = 0 \qquad (4.23)$$

$$\frac{\ddot{a}}{ac^2} + \frac{2\dot{a}^2}{a^2c^2} + \frac{2k}{a^2} = 0 \qquad (4.24)$$

$$\frac{\ddot{a}}{ac^2} + \frac{2\dot{a}^2}{a^2c^2} + \frac{2k}{a^2} = R - \frac{1}{2}(g)R \qquad (4.25)$$

$$= R - \frac{1}{2}(4)R = R - 2R = -R = 0$$

$$R_{\mu\nu} - \frac{1}{(n-1)}g_{\mu\nu}R = 0 \qquad (4.26)$$

Quod Erat Faciendum.

Albert Einstein's' intuition on the geodesic equation of motion was right on the mark.

It is interesting to note that the independent geodesic of motion has been given a general form to a concrete example of gravitation by random large masses that can be derived from the EFEs for empty space. However, the field must not be singular anywhere outside its gravitational points of mass.

The preceding derivation of the EFEs for empty space from the geodesic equation of motion proves that the three dimensionality of time, or the six dimensionality of space-time, are inherent to the General Theory of Relativity, as a complete field theory encompassing fields, particles, and motion, without adding restrictions or external postulates.

Chapter 5

The Quest for Quantum Mechanical Relativity

*§ 1. What are current interpretations of Quantum Mechanics? What
are the quantum mechanical interpretations of "A Dynamic
Theory of Space-Time: A Matter of Waves"?*

Each of the various interpretations of Quantum Mechanics such as,
but not limited to, Copenhagen, Objective Collapse, Retro-Causality,
Super-Determinism, Quantum Bayesianism (QBism), Many Worlds,
Bohmian Mechanics (Pilot Wave), Consciousness Role, Relational,
and Quantum Logic, has merit in the scope of reality of every
wavefunction of our universe.

Nature exists for the divine purpose of its creator, not for the benefit
of an observer's ego. Nature is intrinsically probabilistic, and
efficient. Under the interpretation of A Dynamic Theory of Space-
Time: A Matter of Waves, a Quantum Theory of Gravity, nature may
exhibit the following characteristics:

- The wavefunction is not always collapsible, or self-collapsible,
  at a quantum scale depending on the value of probability.
  Entangled particles may collapse together due to greater
  probability.

- Curvature causes collapse of the wavefunction due to the greater
  superposition of states of the spatiotemporal curvature of
  wavelets.

- Total collapse of the wavefunction everywhere simultaneously
  needs faster than light transfer of information along the paths of
  the particles. Particles may set their initial conditions
  accordingly. It is possible for particles to send information
  backwards in time, which would be equivalent to faster than
  light transfer of information by tachyons.

- The collapse of the wavefunction would be observable at
  classical scales for multiple particles, which is potentially
  testable depending on the state of the art of the testing

technology. When nature was created, it was not based on assumptions for the benefit of a theory.

- The wave function is bi-directional in time, the forward wave is the traditional wavefunction "$\Psi(r)$" and the backward wave is the temporal wavefunction "$\Psi(t)$", this is why time travels at *"c"* through stationary objects of mass.

- The transaction of the forward and backward waves at an instant in the present decides the path to select, the decoherence of the phase of the particle determines the selection of its wave, a force acts on the phase of a particle that is moving faster than light (a tachyon) or backward in time, or on the phase of a particle that is moving slower than light (a tardyon) or forward in time.

- Retro-causality reenforces determinism since all that already happened was reenforced as it was determined. Entanglement is the circuit of super determinism from the big bang to eternity.

- The non-objective probability of Quantum Mechanics is a function of the amplitude to any observer because it is relativistic according to the inherent relativistic geometry of the wave function. An observation is an update on the potential probability of future outcomes.

- Classical probability differs from quantum probability since the wave, phase, and other properties of objects may be different between the quantum and classical worlds.

- Every possible outcome of the wave function of a particle or a system of particles may occur in the universal wave functions of the many worlds that are coincidental or related. It may be possible for universes to collide and join together in the bulk, giving the amalgamation of wave functions greater growth and longevity.

- A particle may be shared efficiently between parallel worlds which makes the spatiotemporal background adjustable from the perspective of the particle. Many worlds may be described as equivalent to many adjustable wavefunctions through the same particle. The split of many worlds may occur due to the multiple probability of divergent paths between the probable directions of the multiple wavefunctions.

- In the absence of any other force, a particle may not move on its own, its phase may be shifted through interactions, and the particle may be carried by a single or a resultant spatiotemporal wavelet (the forward or the backward wave function) at a speed that depends on its mass at any expanding or contracting spatiotemporal point.

- Interactions at every point may be felt instantaneously anywhere through retro-causality and entanglement.

- The role of consciousness in the reality of nature manifests itself efficiently, when a particle or object that is shared by coincidental worlds, is observed, or measured, by the specific manifestation of an observer, or another measuring particle or object, which collapses the probability in its own world because that particle, object, or observer, shares the same wave function and phase as the particle or object being observed or measured.

- Quantum logic affirms that particles communicate forward and backward in time according to predictable probabilities; quantum probability springs from the properties of the wavefunction.

The advent of Quantum Mechanics called for a paradigm shift in logic for classical or quantum scales. Classical logic evolved from the classical scale of our universe.

§ 2. Are the properties of entanglement, particle-wave duality, and General Relativity related within the framework of Quantum Mechanics?

An entangled system is defined to be a system whose classical or quantum state cannot be factored as a product of states of its local constituents; entanglement may also be defined by coincidence correlations. It is interesting to know that classical systems can be entangled. Entanglement between very different objects is possible. Researchers have proven that our macroscopic world is subject to the laws of quantum physics by successfully entangling a millimeter-sized drum, a silicon nitride membrane, with a large cloud of cesium atoms.

The principle of General Relativity and entanglement are functional aspects or attributes of the framework of Quantum Mechanics that scale up to the classical level according to the spatiotemporal wave theory while the complementarity principle for the particle-wave duality quantum aspect changes at the classical level to build up the classical objects or macroscopic particles of the observable physical reality.

If the physical properties of entangled particles like spin, polarization, position, and momentum are measured, these quantum properties can be found to be highly correlated in some cases. Let us consider a system that consist of two quantum particles that are entangled, with position basis of $|x\rangle_1$ and $|-x\rangle_2$ using Dirac deltas, for spatial distances of particles 1 and 2, where x is a variable of a physical property like position, identified by a subscript for either particle 1 or 2.

The positions of the particles are highly correlated, since if particle 1 is at x when we measure particle 2's position, particle 2 would be at $-x$, which shows that the position of particle 2 is highly correlated with the position of particle 1. Consequently, the position state vector is projected onto the direct product $|x\rangle_1|-x\rangle_2$ or $|x\rangle_1 \otimes |-x\rangle_2$, and the center of mass is at zero. When the position of the center of mass of a system is well defined or has zero total angular momentum, even if the positions of the constituent particles are not well defined, the constituent particles can be highly correlated. Then, the state of the entangled system cannot be factored as a product of states of the individual particles.

The equation for the principle of entanglement of the wave-function is given by

$$|\Psi(x)\rangle = \frac{1}{\sqrt{2}}\left(|x\rangle_1 + |-x\rangle_2\right)e^{i\left(\frac{v_r}{c_0}\right)^2} \qquad (2.1)$$

The above equation states in a mathematical way that the quantum state of particle 1 (position 1) cannot be described independently from the quantum state of particle 2 (position 2), and vice versa, even for a very long separation of spatial distance. It is interesting to point out that the spatial distance between the particles may be expanding or contracting as time passes.

The complementarity principle of the particle-wave duality has been verified for elementary particles and compound particles such as atoms or molecules. The wave attribute of a classical object or a macroscopic particle cannot usually be detected because of its extremely short wavelength, its large mass or kinetic energy,

$\lambda_B = h/p = h/\sqrt{2m \cdot (K.E.)}$. To paraphrase Eisberg and Resnick, "both microscopic and macroscopic objects with small or large wavelengths have matter and radiation that exhibit aspects of a particle and a wave. The wave aspects of their motion are more difficult to observe as the wavelengths become shorter. For an ordinary macroscopic object of mass, size, and momentum are always sufficient large to make the de Broglie wavelength small enough to be beyond the range of experimental detection, and classical mechanics reigns supreme." (Eisberg, 1985)

Hence, as the de Broglie wavelength of an object or particle shortens exceedingly compared to its diameter *"D"* or c_0T, the inadequacy of the particle-wave duality of quantum objects and particles at the classical level becomes the adequacy of the classical objects and waves as these attributes become unique concepts to describe the observable elements and aspects of the classical universe. However, the particle-wave attribute is still there as an inactive quantum mechanical characteristic at the classical level.

In quantum mechanics the expression of an overlap of states $\langle p \| w \rangle$ is typically interpreted as the probability amplitude for the state *"w"* to collapse into the state *"p"*. Furthermore, the direct product $|w\rangle|p\rangle$ may be used for describing a composite system, like a particle-wave system.

The complementarity principle of the particle-wave duality may be expressed as

$$|\Psi\rangle = \frac{1}{\sqrt{2}}\left[\langle p||w\rangle + \left(|w\rangle|p\rangle\right)\right] \equiv me^{i\left(\frac{\lambda}{D}\right)^2} \approx m\left[\cos\left(\frac{\lambda}{D}\right)^2 + i\sin\left(\frac{\lambda}{D}\right)^2\right] \quad (2.2)$$

$$\lim_{\lambda \to 0} me^{i\left(\frac{\lambda}{D}\right)^2} = m\left[\cos\left(\frac{\lambda}{D}\right)^2\right] \approx m \quad \to \quad particle \quad\quad (2.3)$$

$$\lim_{\lambda \to \infty} me^{i\left(\frac{\lambda}{D}\right)^2} \approx m\left[\cos\left(\frac{\lambda}{D}\right)^2 + i\sin\left(\frac{\lambda}{D}\right)^2\right] \quad \to \quad particle + wave \quad (2.4)$$

The conformal property preserves both angles and the shapes of the surfaces of infinitesimally small spherical or ellipsoidal waves, but not necessarily their size, curvature, or effectiveness, as the scale of the shapes increases. Therefore, the particle-wave duality, General Relativity, and entanglement are scalable in our physical reality, but their functions are not equally effective because all three quantum mechanical mechanisms are emerging from the spatiotemporal wave function but not all are essential for the interaction of macrosystems.

The General Theory of Relativity, Quantum Mechanics, and their features provide a complementary picture of physical reality, jointly they can fully explain the electromagnetic phenomena of light. In the last one hundred years of Quantum Mechanics, nothing has been lost but time. To paraphrase the eminent physicist Sean M. Carrol during an interview, the gravitizing of Quantum Mechanism Approach is an extreme version of wave function realism. The way forward is to gravitize Quantum Mechanics instead of quantizing gravity. This is a view that says there is no classical world or classical observers. The position and velocity of a particle does not exist in Quantum Mechanics. There are no hidden variables. Nature starts with wave functions not with a classical theory. Space and time are parts of the wave function. The wave function of the universe is objectively real. Everything emerges from the wave function even a Quantum Theory of Gravitation that scales up to the General Theory of Relativity. A spatiotemporal manifold emerges from the space-time of the wave function. Every region of space-time contains a large number of quantum mechanical degrees of freedom that are entangled with each other. Most of our universe is space-time. Distances between different points, near or far, in the space-time of the wave function, may be defined by Quantum Mechanical Entanglement. There is higher entanglement between points that are near, closely related points, and lower entanglement between points that are farther away. Energy can change the entanglement of points in the space-time of the wave function. A quantum description of the wave function quantization is needed for each fundamental particle. (Carrol, 2020)

§ 3. Are the Correlations of Quantum Mechanics due to Spacetime?

If the spins of two entangled particles are measured in two orthogonal directions, the spins would have the same kind of reciprocal uncertainty than position and momentum. According to the uncertainty principle of Quantum Mechanics, if you measure the spin in one direction, you cannot assign an exact simultaneous value for the other spin. Hence, the uncertainty principle of Quantum Mechanics would assert a fundamental limit to the precision, through any of several mathematical inequalities, with which the values for the spins of the two entangled particles in two orthogonal directions can be predicted from initial conditions. So, if one were measuring the spin of one of the particles in the up-down vertical direction, and the other particle in the left-right horizontal direction, there would not be correlation between the simultaneous measurements. However, if the measurements are performed in the same direction, then the measurements would be maximally correlated. If one of the two orthogonal directions of the measurements is adjusted toward a parallel direction, it is possible to determine how strongly correlated the measurement outcomes become if the spins are determined already before particle decay. In such case, the measured correlation would have the upper bound of Bell's inequality. Quantum mechanical experiments have shown that such upper bound can be violated. A difficulty arises if the measurement value for the spin has been determined when the entangled state of the two particles was created, because one could not explain the observed correlations between the spins of the two particles. If the measurement value was not determined at entanglement, then the spins would become determined non-locally on both sides the moment one measures at least one of the spin values. Such a correlation is stronger than it could possibly be if the spin had been determined before the measurement. Thus, these types of measurement results oppose determinism and locality in favor of unpredictability and non-locality.

Let us imagine two entangled particles that travel away from each other, at the same speed of light, in a parallel direction through a spatiotemporal dimension in isotropic and homogeneous spacetime. The advanced temporal waves between them follow parallel paths from their instant localities to their origin of entanglement. Even though the spatiotemporal paths may not be exactly alike in terms of, but not limited to, expansion, contraction, or torsion, it is hypothesized that the advanced temporal waves, as well as its tachyons, traveling between the two entangled particles, would provide the information exchange or handshake effectively, when a measurement is performed on either particle, since the paths are directions of the same dimension. If the paths of the particles were through two orthogonal dimensions, the advanced temporal waves, or tachyons, would not follow their directional paths through the same dimension. It is hypothesized that advanced temporal waves, or the phases of their tachyons, between two entangled particles, may not be equally affected by spatiotemporal perturbations, when traveling through the directions of different dimensions. As a result, a measurement on either particle would be uncertain and the correlation of measurements would be reciprocally uncertain. However, if one of the two orthogonal directions of the measurements is adjusted toward a parallel direction, as one of the directional paths becomes less orthogonal and more parallel to the other, the measurements may become correlated at a lower bound of Bell's inequality since the advanced temporal waves, or the phases of their tachyons, may be less affected by the spatiotemporal perturbations of the direction of the orthogonal dimension. Additionally, if the spins had been determined at entanglement, it is hypothesized that the observed correlations between the spins of the two particles are due to the deterministic conditions of the particles that exists at the locality of origin which are strongly conserved by the advanced temporal waves and their tachyons through the directional paths of the particles. Causality would be preserved when measurement is performed before particle decay. Otherwise, the spins would become determined non-locally on both sides the moment one measures at least one of the spin values. Hence, if these hypotheses are supported, the correlations of Quantum Mechanics and Causality may be directly related to spatiotemporal wave theory.

Chapter 6

Equivalence signifies Equality, and Current Science Topics

§ 1. Why does the wave function follow Schrödinger's equation? Can Einstein's field equations be derived from Schrödinger's equation?

The wave function of Quantum Mechanics guides the quantum particle and tells the location and velocity of the particle. A particle follows the geometry of space-time as it is guided by the spatiotemporal wave function which is represented by Schrodinger's equation in a spatiotemporal medium described by the Einstein's Field Equations of the General Theory of Relativity.

Let us describe the one-dimensional time independent Schrödinger's equation for the wave function.

$$\frac{\partial^2 \Psi}{\partial x^2} + \frac{8\pi^2 m}{h^2}(E - V)\Psi = 0 \qquad (1.1)$$

Where Ψ is the wave function, x is a spatial distance or position, m is mass, h is the Planck constant, E is energy, and V is potential energy.

Defining constants in terms of units and other constants,

$$G \equiv \frac{r^3}{m \cdot s^2} = \frac{r^5}{h \cdot s^3} = \frac{c^3 \cdot r^2}{h} \qquad (1.2)$$

$$\frac{8\pi G}{c^4} \equiv \frac{4 \cdot 2\pi \cdot c^3 \cdot r^2}{hc^4} = \frac{4r^2}{\hbar c} = \frac{4r}{\hbar f} = \frac{4 \cdot 2\pi r}{\hbar \omega} = \frac{8\pi r}{\hbar \omega} \qquad (1.3)$$

$$G = \frac{c^4 r}{\hbar \omega} \qquad (1.4)$$

$$\hbar \omega = \frac{c^4 r}{G} \equiv \text{Force} \cdot \text{distance} \equiv \text{Energy} \qquad (1.5)$$

$$\frac{\hbar\omega}{r} = \frac{c^4}{G} \equiv \text{A force} \tag{1.6}$$

Defining the wave function with Schrödinger's equation over space and time,

$$i\hbar\frac{\partial\Psi(r,t)}{\partial t} = -\frac{\hbar^2}{2m}\nabla^2\Psi(r,t) + V\Psi(r,t) \tag{1.7}$$

$$\text{Total Energy} = -(\text{Kinetic Energy}) + \text{Potential Energy} \tag{1.8}$$

Where $\hbar$ is the reduced Planck constant, "i" is equal to an imaginary number $\sqrt{-1}$, $\Psi(r,t)$ is the wave function defined over space and time, m is mass, ∇^2 is the Laplacian operator for three-dimensional space, and $V(r,t)$ is the potential energy defined over space and time.

Let us define how the total energy minus the potential energy is equal to the kinetic energy as a particle follows the geometry of space-time as described by the Einstein's field equations. Let us imagine that a particle is free falling in a gravitational field following the geometry of the spatiotemporal curvature produced by a celestial body of mass. Hence, if no other force acts on the particle, the change in kinetic energy represents the change in the distance of curvature along its trajectory due to the energy density of the mass, or the proportional spatiotemporal pressure, that produces the gravitational field acting on the particle.

In such scenario, would the Schrödinger's equation for the density of Kinetic Energy be equivalent to the Einstein's field equations of the General Theory of Relativity?

$$i\hbar\frac{\partial\Psi(r,t)}{r^3\partial t} - \frac{V}{r^3}\Psi(r,t)\overset{?}{=}\frac{c^4}{G}G_{\mu\nu} = 4\pi T_{\mu\nu} \tag{1.9}$$

The eminent and humorous physicist Richard Feynman allegedly said "Where did we get Schrodinger's equation from? Nowhere. It is not possible to derive it from anything you know. It came out of the mind of Schrödinger." Let us imagine a journey into the mindset of

Schrödinger to analyze his equation, to shed new light on where the equation may have come from, and to find if it came from the same principles as the General Theory of Relativity. Is it possible that Schrodinger and Einstein were conceptualizing their equations about the same natural phenomenon from opposite ends of the spatiotemporal spectrum of reality, the quantum world versus the classical world? (Hey, 2009)

The density of mass, energy, or matter in space-time is proportional to the spatiotemporal pressure about the geometry of the mass, energy, or matter.

$$\frac{m \cdot r^2}{t^2 \cdot r^3} = \frac{E}{r^3} = \frac{F \cdot r}{r^3} = \frac{F}{r^2} \qquad (1.10)$$

Where "r" is a spatial distance, "t" is a temporal distance, E is energy, m is mass, and F is a force.

Defining the components of the four-dimensional Schrödinger's equation with all three temporal dimensions folded into one and three spatial dimensions,

$$i\hbar \frac{\partial \Psi(r,t)}{\partial t} = i \left[\frac{m \cdot r^2}{2\pi t^2} \right] \Psi \qquad (1.11)$$

$$V\Psi(r,t) = [\hbar\omega]\Psi = \left[\frac{m \cdot r^2}{2\pi t^2} \right] \Psi \qquad (1.12)$$

$$-\frac{\hbar^2}{2m} \nabla^2 \Psi(r,t) = -\left[\frac{m^2 \cdot r^4}{8\pi^2 t^2 \cdot m} \cdot \frac{1}{r^2} \right] \Psi = -\left[\frac{m \cdot r^2}{8\pi^2 t^2} \right] \Psi \quad (1.13)$$

Reassembling all three components of the Schrodinger's equation in terms of variables of mass, space, and time.

$$i \left[\frac{m \cdot r^2}{2\pi t^2} \right] \Psi = -\left[\frac{m \cdot r^2}{8\pi^2 t^2} \right] \Psi + \left[\frac{m \cdot r^2}{2\pi t^2} \right] \Psi \qquad (1.14)$$

Dividing the equation by $1/r^3$ to express energy density or pressure,

$$-i\left[\frac{m\cdot c^2}{2\pi r^3}\right]\Psi = \left[\frac{m\cdot c^2}{8\pi^2 r^3}\right]\Psi - \left[\frac{m\cdot c^2}{2\pi r^3}\right]\Psi \qquad (1.15)$$

Multiplying the previous equation by 2π to simplify terms,

$$-i\left[\frac{m\cdot c^2}{r^3}\right]\Psi = \left[\frac{m\cdot c^2}{4\pi r^3}\right]\Psi - \left[\frac{m\cdot c^2}{r^3}\right]\Psi \qquad (1.16)$$

$$\left[\frac{m\cdot c^2}{r^3}\right]\Psi - i\left[\frac{m\cdot c^2}{r^3}\right]\Psi = \left[\frac{m\cdot c^2}{4\pi r^3}\right]\Psi \qquad (1.17)$$

$$4\pi\left[\left(\frac{E}{r^3}\right)\Psi - i\left(\frac{F}{r^2}\right)\Psi\right] = \left[\frac{F}{r^2}\right]\Psi \qquad (1.18)$$

Multiplying by the trace value "–6" of the Ricci tensor for the curvature according to General Relativity.

$$-24\pi\left[(\rho)\Psi - i(p)\Psi\right] = \left[(-6)\left(\frac{c^4}{Gr^2}\right)\right]\Psi \qquad (1.19)$$

In the above equation "i" is equal to a rotation of 90^0 because reactive pressure is applied perpendicularly by the boundary of mass, energy, or matter. Only the perpendicular component of pressure does work on the boundary of the spherical geometry of the object. The term $(1/r^2)$ represents curvature, ρ is the mass density and p is the pressure from the stress-energy-momentum tensor $\left(T = g^{\mu\nu}T_{\mu\nu} = -3\rho + 3p\right)$.

Aside, from previous research, the coefficient of the energy density and the pressure variable is "3" if time is unfolded into its three temporal dimensions. Each spatial dimension has a conjugate temporal dimension. Space-time is complex. (Nieves, 2020)

$$-\frac{8\pi G}{c^4}\left[\left(-3\rho\right)\Psi+\left(3p\angle 90^\circ\right)\Psi\right]=\left[\left(-6\right)\left(\frac{1}{r^2}\right)\right]\Psi \qquad (1.20)$$

Rearranging the above equation to a more recognizable Einstein's field equation, using only the magnitude of pressure, we have

$$-6\left(\frac{1}{r^2}\right)\Psi=-\frac{8\pi G}{c^4}\left(-3\rho+3p\right)\Psi \qquad (1.21)$$

Replacing the variable of curvature $(1/r^2)$ by the trace R of the Ricci curvature tensor, where n is equal to four dimensions as in the original equation of Einstein, we obtain

$$\left(-R\right)\Psi=-\frac{4\pi G}{c^4}\left(-\rho+p\right)\Psi \qquad (1.22)$$

$$\left(R_{\mu\nu}-\frac{1}{\left(n-1\right)}g_{\mu\nu}R\right)\Psi=\frac{4\pi G}{c^4}T_{\mu\nu}\Psi \qquad (1.23)$$

$$\left(R_{\mu\nu}-\frac{1}{3}g_{\mu\nu}R\right)\Psi=\frac{4\pi G}{c^4}T_{\mu\nu}\Psi \qquad (1.24)$$

Dividing both sides of the equation by the wave function Ψ to leave only the Einstein curvature tensor and the stress-energy-momentum tensor of the General Theory of Relativity,

$$G_{\mu\nu}=\frac{4\pi G}{c^4}T_{\mu\nu} \qquad (1.25)$$

Quod Erat Demonstrantum, the field equations of the General Theory of Relativity emerge from Schrodinger's equation that guides the spatiotemporal wave function of Quantum Mechanics.

§ 2. What is the six-dimensional equation for Total Energy?

Kinematics is the field of physics that describes the motion of the bodies and finds out velocities or accelerations for various objects. On the other hand, Kinetics describes how a body reacts when a force or a torque is applied to it. Since the total energy equation is a kinetic equation, the total energy is potentially kinetic in nature, and may be expressed mathematically as such, or as a total momentum equation for a totally elastic instantaneous collision.

$$\text{Total Energy} = -\,(\text{Kinetic Energy}) + \text{Potential Energy} \qquad (2.1)$$

$$\frac{mc^2}{8\pi} = -\frac{1}{8\pi}\frac{h^2c}{m\lambda^2} + \pi mgrc \qquad (2.2)$$

Converting the equation from 4 to 6 dimensions by multiplying by "c" to denote the six-dimensional total energy equation,

$$\frac{mc^3}{4\pi} = -\frac{h^2c}{4\pi m\lambda^2} + 2\pi hg \qquad (2.3)$$

$$i\frac{m^2G}{4\pi}\frac{\partial \Psi(r,t)}{\partial t} = -\frac{m^2G\omega}{4\pi}\nabla^2\Psi(r,t) + 2\pi cV\Psi(r,t) \qquad (2.4)$$

$$i\frac{\hbar^2}{mr}\frac{\partial \Psi(r,t)}{\partial t} = -\frac{\hbar^2\omega}{mr}\nabla^2\Psi(r,t) + 2cV\Psi(r,t) \qquad (2.5)$$

$$i\hbar^2\frac{\partial \Psi(r,t)}{\partial t} = -\hbar^2\omega\nabla^2\Psi(r,t) + 2hV\Psi(r,t) \qquad (2.6)$$

The momentum-energy or "momenergy" of a particle traveling through six-dimensional space-time may be expressed in its simplest terms as

$$m^2c^3 = -\frac{h^2c}{\lambda^2} + 2m^2grc \qquad (2.7)$$

$$p\cdot mc^2 = -p^2c + 2p\cdot mc^2 \qquad (2.8)$$

Therefore, it is interesting to note that six-dimensional Einsteinian energy times mass is momenergy.

Since six-dimensional space-time has three spatial dimensions and three temporal dimensions, the product of mass squared with a spatiotemporal displacement, that consists of three spatial components and three temporal components, may be combined into one quantity called momenergy. The magnitude of the momenergy vector is a particular combination of energy and momentum that is invariant.

Thus, let us denote momenergy as follows:

$$Momenergy\ (p \cdot mc^2) = (mass)^2 \times \frac{Spatial\ Displacement\ (m^3)}{Temporal\ Displacement\ (s^3)} \quad (2.9)$$

The direction of the momenergy of the particle is the direction of the particle's worldline at a specific instant of time. Momentum and energy are relativistic and conserved. *Energy is proportional to momentum in a similar way that space is proportional to time.*

§ 3. The spatiotemporal rotating distortion of two charges.

A spatiotemporal bridge is thought to be a cosmic connection between two regions of space-time within one universe or between two universes. Theoretically, a spatiotemporal bridge may help a signal to travel between two spatiotemporal locations that are very far apart, by providing a shorter path between an emitter at the sending point to a receiver at the receiving point in space-time, if that were possible through a Lorentz bridge, than if the signal were to travel through the universe at the speed of light. A spatiotemporal bridge is mathematically possible and begs the question: Can it be scientifically demonstrated?

The General Theory of Relativity allows for time travel to the future, but not the past due to causality, unless the technology allows for travel faster than light to parallel universes or other scenarios that have been considered. Spatiotemporal bridges are accepted solutions to the EFEs.

In 1916 Ludwig Flamm realized that in certain coordinate systems, the gravitational hole described by the Schwarzschild solution to the EFEs was a two-sided funnel or a spatiotemporal bridge, which connects a black hole to a white hole. The entrance, or black hole, and the exit, or white hole, could be in the same universe or in parallel universes. The Schwarzschild solution describes two symmetrical spatiotemporal regions, and the two-sided funnel is the cosmic connection between them.

In 1935, Albert Einstein and Nathan Rosen expanded Flamm's realization to a theory of particles. They imagined two spatiotemporal bridges in two spatiotemporal regions within the same universe. On these spatiotemporal regions the bridges connecting the two regions would behave like two particles. These particles could move around and interact with each other. These bridges could be thread with electromagnetic field lines, and they would behave like interacting charged particles. (Einstein, 1935)

Let us further imagine that these bridges are oppositely charged as in a dipole, with one positive and the other negative, surrounded by an electromagnetic field. These Einstein-Rosen bridges would behave similar to an electron-positron pair. This bridge would allow for quantum time travel. (Fuller, 1962) If the entrance end of a traversable bridge is stationary and the exit end is rotated at the speed of light, then theoretically, the time traveler could always travel back in time to the instant where the exit end was accelerated and time was not passing, but not farther back in time when the bridge did not exist yet. What would happen to causality then? If the Schwarzschild solution is used, for a non-rotating bridge, the bridge would collapse so rapidly that not even a signal may traverse it. Causality is not violated.

The EFEs allow for any topology and smoothly changing geometry of the spatiotemporal fabric. It is currently thought that the only limitation of space-time has to do with the nature of energy and matter that space-time may contain. Spatiotemporal geometry defines the distribution of matter, or energy, for example, to keep the spatiotemporal bridge open. The exotic energy may be an electromagnetic field and a field potential. These conditions would not violate the allowable distribution constraints of the EFEs.

Let us consider an advanced technology that allows a hyperdrive to produce a rotating electron-positron pair, or a rotating electron-electron pair, to initiate a spatiotemporal distortion, and two spatiotemporal bridges are created that could be thread with electromagnetic field lines and a field potential, where these bridges would behave like interacting charged particles. (Alcubierre, 1994) If the hyperdrive uses an electron-electron pair, it is theoretically possible to hypothesize that voltage potential, a magnetic field, and torsion, may be used to tread, to widen the opening, and to maintain the stability of the spatiotemporal bridge in its entire timeline. A signal may travel through the spatiotemporal bridge from an emitter to a receiver that are far apart in the same universe. Causality is not violated in a quantum mechanical many-worlds scenario. (Everett, 1973)

From previous theoretical research, a spatial bridge between two regions of the universe may provide a faster passage through space. So, the Category-one spatiotemporal bridge would be mostly spatial, with temporal coordinate changes that are almost negligible. Both gates (portals) on the bridge would be spatially separated, but nearly simultaneous on the temporal medium. A Category-two spatiotemporal bridge would be both spatial and temporal, but the change in temporal coordinates are toward the future. A Category-three bridge is a spatial, temporal, and interdimensional bridge through six-dimensional space-time. If the gates are concentric, the temporal distortion effect is greater than the spatial effect on the same spatiotemporal plane, toward the past. Hence, a spatiotemporal bridge may be directed, and later re-directed, aligned, or later re-aligned, in the direction of the resultant spatiotemporal metric toward the future, present, or past. (Nieves, 2020)

Let us imagine a fantastic exploration journey that the young and talented science fiction writer George H. White may have envisioned about an interdimensional probe with artificial intelligence and other advanced technology that may be sent to the past of the earth through a spatiotemporal bridge to record and archive historical events of humanity and geology from an earth orbit. (White, 1978) The astronomical or magnetic alignment or orientation of structures such as temples, pyramids, tombs, stonehenges, and other monuments, may be used as chronological clocks for proper earth time by the probe to update or verify the accuracy of its destination clock and recordings. Massive durable structures may be constructed during a pre-civilization period or an early civilization period to serve as chronological landmarks for later exploration. It would be important that enough of these massive structures would remain as a legacy of engineering, science, and architecture for any future civilization.

To paraphrase Albert Einstein, "according to Hapgood, the virtually rigid outer crust of the earth undergoes, from time to time, an extensive displacement over the inner layers". Moreover, it has been long understood that the magnetic poles of the earth (dip poles) migrate over time, or that the magnetic poles could flip every 200,000 to 300,000 years. The geological studies of the probe may help to correct inaccuracies as well as confirm these theories, to improve the accuracy of the tracking of the magnetic poles of the earth for navigation and other purposes. (Hapgood, 1958) (Brown, 1967)

By tracking the magnetic pole (the sink) on the geographical north pole of the earth where magnetic lines converge, an exploration probe may calculate a more accurate proper earth time in the past. The WWV is the radio call sign for the National Institute of Standards and Technology's time and frequency shortwave radio station. It is the oldest continuously operating radio station in the United States.

Since 1920, the WWV has been sending, from different transmitters, several high frequency signals in the radio spectrum between 5-20 MHz, that are accurate to 0.0001 milliseconds. These transmitters were relocated to Maryland sites in 1931, and then to Fort Collins, Colorado, in 1966. The exploration probe could easily use these signals for accurate proper earth time. Since the proliferation of the internet in the 1990s, the World Wide Web, a descendant of ARPANET, a backbone interconnection of academic and military networks from the 1970s, may also become a crucial source of historical information for the probe's archive.

After envisioning his fantastic exploration journey for a while, the talented George H. White would have broken his reverie when he heard his wife say "darling, your elevenses will get cold!"

§ 4. What would be the benefit of synthesizing heavy poor metals?

The element Ununpentium was first synthesized in 2003 by a joint team of Russian and American scientists at the Joint Institute for Nuclear Research (JINR) in Dubna, Russia. In December 2015, it was recognized as one of four new elements by the Joint Working Party of international scientific bodies IUPAC and IUPAP. Ununpentium is an extremely radioactive element, its most stable known isotope, Unupentium-290, has a half-life of only 0.65 seconds.

No other properties, other than nuclear properties, of Ununpentium or its compounds have been measured, because of its extremely limited and expensive production and its fast decay.

Properties of Ununpentium remain unknown and only predictions are available, which begs the question: what would be the benefit of synthesizing a heavy poor metal like Ununpentium? (Subramanian, 2019)

Let us perform a gravitational analysis of a body of mass of Ununpentium. Let us look into the gravitational acceleration of a body of mass that consists of a heavy poor metal to analyze how gravitation behaves at the atomic level and at a large-scale level.

Let us look into one of the elements that has been recently manufactured synthetically in the lab. The element Ununpentium has the characteristics of a heavy poor metal as shown below. (St. Fleur, 2016)

$$Surface\ Area\ (403\ particles\,/\,atom) = \frac{4\pi\left(ar_e^2 + br_p^2 + cr_n^2\right)}{1\ atom} \quad (4.1)$$

Where a, b, and c are the number of electrons, protons, and neutrons per atom of Uup.

$$Number\ of\ Atoms = \frac{1\ Kg\ of\ Uup}{0.288\ Kg\,/\,mole} \ x\ atoms\,/\,mole \quad (4.2)$$

$$Avogadro's\ Number = atoms\,/\,mole$$

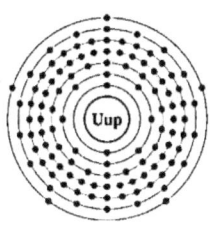

Figure 1. Ununpentium Electron Configuration

Ununpentium	Uup
Configuration (Predicted)	$1s^2 2s^2 2p^6 3s^2 3p^6 4s^2 3d^{10} 4p^6 5s^2 4d^{10} 5p^6 6s^2 4f^{14} 5d^{10} 6p^6 7s^2 5f^{14} 6d^{10} 7p^3$
Electrons/Shell (Predicted)	2, 8, 18, 32, 32, 18, 5
Velocity of Electrons	$\sim c$
Melting Point (Predicted)	$400\,^{0}C$
Boiling Point (Predicted)	$\sim 1100\,^{0}C$
Atomic Radius (Empirical)	$r_a = 187 \times 10^{-12} m$
Number of Electrons (a)	115 (28.54%)
Radius (1e)	Classical: $2.8179403227 \times 10^{-15} m$ Observational: $10^{-18} m - 10^{-22} m$ (used $10^{-20} m$)
Surface Area (115 e's)	$115 \times 4\pi \times 10^{-40} m^2 = 1.445132621 \times 10^{-36} m^2$
Number of Protons (b)	115 (28.54%)
Radius (1p)	$0.84 - 0.87 \times 10^{-15} m$ (used $0.855 \times 10^{-15} m$)
Surface Area (115 p's)	$1.056428074 \times 10^{-27} m^2$
Number of Neutrons (n)	173 (42.92%)
Radius (1n)	$0.8 \times 10^{-15} m$
Surface Area (173 n's)	$1.391348554 \times 10^{-27} m^2$
Total Surface Area (403)	$2.447776629 \times 10^{-27} m^2$
Total Number of Particles/Isotope	403 (100%)
Atomic Mass	288 u (amu)
Weight/Atom (403)	Calculated from weights of 403 atoms: $4.821356048 \times 10^{-25} Kg$ Calculated from 288 u: $4.782352435 \times 10^{-25} Kg$ 1 amu (1 u) $\approx 1.6605390420 \times 10^{-27} Kg$

Density (Predicted)	$13.5\ g/cm^3 = 13500\ Kg/m^3$
Alpha Radius (1 Atom)	$r_a = \sqrt[3]{\text{Total Surface Area } (403)/4\pi}$ $r_a \approx \sqrt[3]{1.94787875 \times 10^{-28} m^2} = 1.395664269 \times 10^{-4}\ m$
Avogadro's Number	$6.022 \times 10^{23}\ atoms/mole$
Number of Moles in $1Kg$	$3.472303439\ moles$ (From 288 u) $0.003444213\ moles$ (From sum of atomic weights)
Number of Atoms in $1Kg$ (N)	$2.091021131 \times 10^{24}\ atoms/Kg$ (From 288 u) $2.074105273 \times 10^{21}\ atoms/Kg$ (From sum of atomic weights)
Alpha Surface Area (All atoms in $1Kg$)	$0.005118353\ m^2$ (From 288 u) $0.000005077\ m^2$ (From sum of atomic weights)
Beta Radius $(1Kg/m^3)$	$r_\beta \approx 0.05588464\ m$
Beta Surface Area $(1Kg/m^3)$	$0.008529586\ m^2\ for\ sphere$
Alpha/Beta Surface Area Ratio	$4\pi r_a^2/4\pi r_\beta^2 = r_a^2/r_\beta^2 \approx 0.60007050 8$ (From 288 u)
Alpha Radius/Beta Radius	$r_a/r_\beta \approx 0.774642181$ (From 288 u)
Alpha Radius/Atomic Radius	$r_a/r_A \approx 0.00007 4634 \approx \sqrt[3]{6} \times 10^{-7}$
Earth's Standard Gravity (g)	$9.80655\ m/s^2$
Beta Gravity $(1Kg)$	$g_\beta \propto \dfrac{r_a^2}{r_\beta^2} g_a \qquad \dfrac{g_\beta}{g_a} \propto \dfrac{r_a^2}{r_\beta^2}$
Alpha Gravity $(1Kg)$	$g_a \propto \dfrac{r_\beta^2}{r_a^2} g_\beta$

Figure 2. Characteristics of the element Ununpentium.

The atomic radius is a measure of the size of the radius of a single atom of an element, usually the mean or typical distance from the center of the nucleus to a spherical boundary of the surrounding cloud of electrons. The boundary is not a well-defined physical entity. The electrons of Ununpentium move at velocities comparable to the velocity of light. The atomic mass is approximately equal to the number of protons and neutrons in an atom, or the average number allowing for the relative abundances of different isotopes.

The heavy poor metal Ununpentium, and lower elements like the lanthanides (rare earth elements) have atoms with shorter radii than would have been previously predicted, due to heavy poor metal shielding or lanthanide shielding. The heavy poor metal shielding has to do with the electromagnetic attraction between the nucleus and the outermost electrons which are partly shielded by the inner electrons in the way. The electromagnetic attraction is shielded collectively as more electrons are in the "d" and "f" orbitals. But as the outermost electrons increase for heavier poor metals, the shielding effect diminishes, and the poor metal atomic radii become shorter. As the atomic radius become shorter, the alpha gravitational field of an atom of a heavy poor metal like Ununpentium extends farther than its radius.

The inner electrons of the heavy poor metal elements move at relativistic speeds as the charge and the particles of the nucleus increase which enhances the electromagnetic effect on the outermost electrons. These relativistic electrons are heavier and further shorten the atomic radius of the poor metal element, also affecting the orbitals of electrons in a complex way. Research on Copernicium 112, has provided experimental evidence of these characteristics for heavy poor metals and lanthanides. (Schwerdtfeger et Al, 2008)

It has been possible to calculate that the empirical atomic radius r_A extends past the infinitesimal perimeter of the gravitational radius r_α of an atom of Ununpentium, allowing for a greater spatiotemporal curvature potential at the atomic level. Hence, it is possible to suggest that the Alpha G-waves may be accessible outside of the gravitational radius r_α of the Ununpentium atom. The Alpha G-wave has wavelength, frequency, and amplitude like any other wave that may be modulated. The greater Alpha spatiotemporal curvature $\left(1/m^2\right)$ of all the atoms in 1 Kg of Ununpentium produces greater gravitational acceleration, than the Beta spatiotemporal curvature for the same quantity of mass, that is measurable at a large-scale of matter outside of the body of mass. Consequently, the gravitational acceleration of a single atom of Ununpentium, the Alpha G-wave, is less than the gravitational acceleration of the large-scale body of mass, but the summation of Alpha G-waves of all atoms of the same quantity of mass is greater than the Beta G-waves of the large-scale body of mass.

If the volumetric acceleration of space-time at the gravitational radius r_α is proportional to the volumetric acceleration at the gravitational radius r_β to conserve gravitational energy of the entire body of mass, we would have

$$g_\alpha r_\alpha^{\,2} \propto g_\beta r_\beta^{\,2} \tag{4.3}$$

$$\frac{g_\alpha}{g_\beta} \propto \frac{r_\beta^{\,2}}{r_\alpha^{\,2}} \tag{4.4}$$

Therefore, in the case that the alpha-to-beta ratio of an atom is 0.774642181 percent, the alpha curvature is approximately 1.290918601, or roughly 1/3 greater, that the beta curvature, producing a greater alpha gravitational acceleration of approximately 1.67, or ~ 5/3, at the integral atomic level.

$$g_\alpha \propto \frac{r_\beta^{\,2}}{r_\alpha^{\,2}} g_\beta \tag{4.5}$$

In the above case, since the beta curvature is less than the alpha curvature, the gravitational acceleration of the Beta G-waves would be approximately 60% to the gravitational acceleration of the Alpha G-waves.

$$g_\beta \propto \frac{r_\alpha^{\,2}}{r_\beta^{\,2}} g_\alpha \tag{4.6}$$

How would a brilliant and competent engineer be able to utilize this heavy poor metal for propulsion?

Now that we have discussed a heavy poor metal element, let us imagine the fantastically popular dream of children in the nineteen fifties and sixties of a flying car, if the dream of building such a vehicle were possible through the technology of gravitational modulation. This science fictional idea of our theoretical Jules-Verne project, was part of the world long ago where children saw in at least one publication that 'the G-engines are coming!' So, let us go ahead and try to fulfill that fanciful dream within our theoretical project.

The connected and autonomous flying car would have four small gravitational projectors on each corner of the underside of the vehicle for directional steering, and a larger gravitational projector in the center of the undercarriage for uplift. Any pair of small diagonal G-projectors, or the largest G-projector at the center, would suffice for complete uplift and spatiotemporal buoyancy. If such a science fictional idea were possible, it would require some kind of gravitational engine that produces what we would call the Alpha G-waves that interfere, or cancel, some of the Beta G-waves emitted by a celestial body of mass like the Earth.

In the gracious style of the eminent engineer Nicola Tesla, the flying car may run on the ground as an electric vehicle for short distances. Each side-walled non-pneumatic tire of the four-wheel drive of the flying electric car would serve as landing gear, and each would have its own electric motor incorporated in the rim of the wheel. For an enthusiast that is more a pilot than a driver the fantastic flying vehicle may be built as a triangular shaped craft without a Nicola Tesla-designed electric motor with only three small gravitational projectors at each corner and a larger gravitational projector in the center.

How would have Tesla designed a small Linear Zapping Reactor and power system of such a fantastic vehicle? Tesla would have probably visualized and realized the design through his renown problem solving skills. The G-engine of the flying electric car would utilize a shielded LZR to bombard the Ununpentium with a particle accelerator in a modulated way that produces the desired quantity of pulsed Alpha G-waves, alpha particle containment, and radiation. The radiation may be utilized, through a thermoelectric converter, to produce electrical power for the batteries and devices on board of the fantastic flying electric car. The particle accelerator injects alpha particles, positive ions of Helium, stripping one electron, to slow acceleration in order to better modulate the beam energy into the tuned and shielded vacuum chamber of the LZR. Four atoms of Ununpentium have been synthesized in 2003 by bombarding

Americium-243 with Calcium-48 ions. These atoms decayed by emission of alpha-particles to Nihonium in about 100 milliseconds. It is imaginable that with such Tesla technology, the Ununpentium may transmute, gaining, or boosting, its number of nucleons to the next heavier element, livermorium, causing an Alpha G-wave phase change to a stronger Alpha G-wave that extends beyond the perimeter of each boosting atom of Ununpentium. Consequently, the Alpha G-waves which are orders of magnitude stronger than the Beta G-waves may be accessible with the right technology.

Mr. Tesla, our brilliant engineer, would think of practical ways to access and guide those Alpha G-waves, through tuned tubes and cavity resonators, to tune and amplify the Alpha G-waves on their way to the gravitational projectors or emitters. The Alpha G-waves would be pulsed at or close to 180 degrees out of phase to the Beta G-waves to interfere and reduce the gravitational acceleration of the planet, or heavier body of mass, to produce levitation of the fantastic flying electric car. Motion in the forward, backward, or lateral direction would be achieved by orienting the steering projectors forward, backward, or to either side, creating a spatiotemporal distortion in the direction of movement. The spatiotemporal distortion, or warping, creates a gravitational differential that propels the flying electric car in the intended direction. Breaking would be achieved by the attenuation, or reversal, of the spatiotemporal distortion, if there is only downward spatiotemporal distortion towards the Earth, the flying electric car would come to a stop and levitate, and possibly slowly rotate, in its place of buoyancy.

Tesla's remaining rhetorical question would be: how could this Ununpentium material be manufactured in large quantities to be a stable slow-gaining isotope for our fanciful project? Relentless as always, Mr. Tesla would probably say jokingly 'hopefully, other brilliant people are working on the problem.'

§ 5. *Experiencing gravitational field effects.*

Let us consider the flame of a Zippo lighter that stands in a small-scale gravitational field that is embedded in a large-scale gravitational field, such as the gravitational field of the earth. As the spatiotemporal waves of the small gravitational field, the Alpha field, emanate from its source, temporal dimensions would contract and spatial dimensions would extend, equal and opposite to the spatiotemporal waves of the large-scale gravitational field, or Beta field. Hence, any of the spatiotemporal waves emanating close to the Zippo lighter interfere and cancel with each spatiotemporal wave from the Beta field. Each of the photonic waves emanating from the source of light surf on its corresponding spatiotemporal wave, so the photonic wave becomes a standing six-dimensional photonic wave. A nearby observer in the Beta field, that observes the flame of a Zippo lighter in the Alpha field, would observe what seems to be a still three-dimensional image of a Zippo lighter with a flame that never wavers. If the Alpha field were interrupted, the Zippo lighter and flame would again appear as they normally would to an observer in the large-scale gravitational Beta field.

Let us now imagine that a man's favorite niece has invited her relatives to her kidney shaped pool, that has two water jets, one the deep end and one on the shallow end, to circulate the water counterclockwise and any potential floating debris into the skimmer. There are several different types of colorful beach balls floating on the water of the pool which move around from the strongest water jet current towards the weakest water jet current. The stronger water waves overcome the weaker water waves at great distances, water waves and gravitational waves follow the same principles. If an observer near the weaker water jet would place his hand on the water jet as the hand moves closer to the water source, the force opposing the movement of the hand would grow in strength to a maximum pressure that would be very hard to overcome. As the hands of the observer, or the body of the observer, move farther from the weaker water jet, the observer starts experiencing the stronger waves of the water jet from the deep end and less the weaker waves from the water jet from the shallow end of the pool.

Consequently, if it were possible for an observer in a large-scale gravitational field, a Beta field, to move his hands about and closer to a small-scale gravitational field, an Alpha field, embedded in the Beta field, the opposing force of the Alpha field would become extremely difficult to counteract by the hands of the observer. Moreover, if the observer threw a baseball at the source of the gravitational Alpha field, the ball would approach the Alpha field up to a point where its kinetic force would equal the force of the field, as it fell further into the field, only to be propelled through space-time by the increasing counter force of the field in an opposite direction.

§ 6. The effect on a corpuscle of pure energy after reaching and exceeding the speed of light.

Let us imagine that a physical or metaphysical corpuscle of energy is able to reach the speed of light by traveling more through space and less through time in isotropic and homogeneous space-time in the direction of the retarded spatiotemporal wave. If an imaginary observer were able to pilot such a corpuscle of energy, the initial cosmic image seen by the observer of a beautiful red nebula would change, as the corpuscle accelerates toward light speed, into an image of the light rays, from such nebula, separately stretching as they pass by. As the corpuscle reaches the speed of light it may be considered a tachyon with imaginary mass.

If a corpuscle of energy consists of quanta of energy of its imaginary mass, each quantum of energy will slow down through time until each quantum reached the speed of light and stops moving through time, while continuing to move at the speed of light through space. However, if there is additional linear momentum energy on each quantum of energy of each corpuscle, the corpuscle may reverse direction and travel in the direction of the advanced spatiotemporal wave where it would utilize the remaining momentum energy to backtrack its previous path if it follows the same path back to its point of departure.

The leading quanta of energy may experience a reorganization of its quanta as a mirror image of its corpuscle as each quantum travels on the advanced spatiotemporal wave in the direction of its point of departure. The arrangement of the quanta of energy of the corpuscle would remain the same in the opposite direction of travel. As if a photon were reflected back by an ideal mirror. This directional change may also be a dimensional change while traveling back in time at a speed greater than the speed of light from the perspective of an observer at an inertial frame of reference traveling much slower than light in the previous direction of the corpuscle in the retarded spatiotemporal wave.

§ 7. What is the field effect of a warp drive on a metallic surface?

Nowadays, there are scientists working on how to engineer a warp drive engine for a spaceship. A force field is a staple of warp drives. Not long ago, a warp drive was a staple of science fiction. If this fantastic technology were to exist, what would be the field effect of a warp drive on a steel surface?

The structure of a solid metal consists of a regular arrangement of closely packed metal ions to form a lattice structure. The valence electrons surrounding the nucleus are constantly moving which makes the metal a good electrical conductor. The electrons of metal atoms leave the outer shell to form a cloud of delocalized electrons. The fluid delocalized electrons allow the metal to bend without breaking. There are several crystal structures of metals, some are cubic or hexagonal. The atoms of a metal are arranged in layers. If a force field were applied to those layers, layers may slide over each other or bend, depending on the strength of the material.

The Elastic Modulus of a material is a fundamental property of every material that is dependent upon temperature and pressure. It is the stiffness of a material or how easily it would bend or stretch under stress. Even at a specific outdoor temperature, it is possible to apply enough pressure on steel, either stress or tensile force per unit of area, to bend it or stretch it. For a certain amount of stress, the strain produced in rubber is larger than the strain produced in steel. A very tough steel may bend or stretch and return to its original shape, unless it is stressed or pulled beyond its yield point, its point of no recovery. It is possible to bend steel using a torch to heat and soften the metal which would ruin any paint on the steel. Thus, according to the Elastic Modulus of steel, it is possible to deduce that steel has greater elasticity than rubber, but a force of significant magnitude may deform it permanently at room temperature.

As the space-time around a steel surface is deformed by a force field in the gravitational field of the earth, a Beta gravitational field, the plenum of the atomic structures of steel is also curved, and an alternating counter gravitational field, or Alpha field, may result. Any object within the Alpha counter gravitational field may feel the alternating part of the overall force field to some extent. If the Alpha gravitational field exceeds the yield point of the steel surface either in an upward or downward direction, the steel surface may be deformed in an undulated fashion in the direction of propagation of the source of the Alpha gravitational field. If the steel surface is moving parallel to the source of the Alpha gravitational field, the undulated deformation would be fast and very loud as well as any other spatiotemporal effect through the region of motion. The undulated deformation of the steel surface would resemble the undulation of the force field. The atomic structure of steel would be pulled and pushed, stretching, or compressing, beyond the yield point of the material depending on the position of the source of the force field with respect to the orientation of the surface.

Most passenger vehicles or cars in the 1940s and 1950s designed for mass production were made of either steel or aluminum. Steel was less expensive than aluminum. Aluminum is lighter than steel and does not rust. Expensive luxury cars or high performance cars were made of aluminum. Steel is still a major component in passenger vehicles, but in the 1940s, the nearly entire frame and chassis of the vehicle were made of unit steel. Passenger vehicles were made sturdier, heavier, and less fuel efficient.

Passenger vehicles of the period were also designed along certain artful appeal with a number of decorative features that are seldom seen in today's vehicles. Those features included chrome highlights and wooden paneling.

Plastic or carbon fiber are being used more frequently in the modern mass production of passenger vehicles. These materials provide a lighter body, better fuel efficiency, and recycling when the vehicles come to the end of their life cycle. Plastic is also easier to work with and to repair at a lower cost. Carbon fiber is very light and strong, but very expensive. It is primarily used on high performance vehicles.

Let us imagine what will happen to an automobile from the 1940s if the driver encounters something unexplainable like the spatiotemporal effect of a nearby warp drive. As a young woman Vanessa Quintus, Victoria's mother, was driving at 64 to 72 kilometers per hour (40 - 45 miles per hour), in the middle lane, along the new M-30 motorway back in 1974 when her classic American car, an imported refurbished black 1947 Ford Super

Deluxe Coupe, suddenly lurched to the right for no apparent reason as her windshield glass gradually shattered. Vanessa did not see any cars within sight at that hour of the moonless night of a weekday in the newly inaugurated motorway. The M-30 orbital motorway circles the central districts of Madrid, the capital city of Spain. It is a ring road of the Spanish city, with a length of 32.5 km (20 miles). She did not see any pedestrians or animals near the car as she drove down the motorway. She immediately pulled over to the side, left her car running with the lights on so other cars could see it, and got out, and was shocked by the appearance of what she saw on the left side of her car.

Even though she was in that moment the only driver on that section of the road, she experienced loud bangs on her car, as if a force were trying to ram her car off the middle lane of the road. At one point the blast was so hard that it trusted her car to the right lane. But the banging stopped as sudden as it had started. There was not a single trace of another vehicle on the dented surfaces that could have happened if the car had been physically pushed.

After she got her bearings, Vanessa stopped the car and rang her husband from a nearby telephone to drive her to the nearest police station to file an accident report and to have the car brought to the authorities. The police officer assigned to her case investigated the damage to the car, could not provide an acceptable explanation for the strange damage on the side of the car. The police investigator noted that there were no scratches on the paint, blood, feathers, or animal hairs, no evidence, on the damaged area of the car. The area looked undulated as if the steel of the body of the car had been melted by a very intense heat blast, or some kind of a force field, but it did not appear burned in any way. The whole left side of the car appears twisted, but no paint was removed at all. The windshield frame was bent enough that it exerted enough pressure to shatter the glass, but nothing had hit the windshield.

Neither the policeman nor her husband could come up with an explanation of how the very extensive damage could have occurred. How was that possible? Vanessa felt a thrust on the left side of her car that made the car shift lanes to the right, and vibration as if the car were shaken or lifted from the road. She held the steering wheel firmly, not swerving out of the lane, and gained control back rapidly.

Then, got out and saw several big undulated dents, some were larger and deeper than others, like a wave. Parts of the car body like the headlight chrome plated lens trim, the left side of the hood of the car, the left fender, the left side of the roof of the car, the windshield frame, were melted and undulated. But there were no cars around before or after the hit and there were also no scratches on the paint of her car from other cars. What unseen force could do this? Why was everything undulated?

This was an American car with a body or chassis made of unit steel with a curb weight of 1465 kg (3230 lbs.). This Ford was made to last, from a time when a car did not typically have plastic or carbon fiber parts on the body. Her husband had lived in the United States and loved classic American cars. His American friends used to joke with him and laugh that he could have a Ford in any color he wanted as long as it was black. The car had a 239/100 HP 3.9L flathead V8 L-Head cast-iron engine, a 3-speed manual transmission, a maroon interior, very original throughout, all new interior, engine completely rebuilt, chrome plated garnish moldings, absolutely immaculate paint, previously owned by a senior diplomat at the U.S. embassy in Madrid who had been reassigned by the U.S. government to another country.

The police investigator told them that he had heard about a very strange incident that happened in northern Spain where a ball of light that was hovering above a rural highway had hit a police patrol car. At a quarter to 2 a.m. in the early 1970s, police officer José M. de los Santos, was on night patrol along a rural section of national road N-120, known as Estrada Logroño Vigo, near Monforte de Lemos in the province of Galicia when he drove into a ball of white light. The N-120 is a Spanish national road that connects the cities of Logroño and Vigo, ending at the port of the latter in northeastern Spain.

José noticed a very bright brilliant light, or ball of energy, 20 to 30 centimeters (8 to 12 inches) in diameter, hovering a meter or a meter and a half (3 to 4 feet) off the ground, as he wrote in his incident report and later explained to a local reporter during an interview. The contour of the ball of light was very well defined and it was not ball lightning. It was a clear and balmy night without thunderstorms or rain. Jose had researched ball lighting and what he saw did not agree with the hypothetical description associated with ball lightning. Ball

110

lightning has not been demonstrated or duplicated satisfactorily in the lab, even today, to show that ball lightning can hover in place above the ground for long periods of time, minutes, or hours, without discharging into the atmosphere within seconds or exploding in mid-air. Though ball lightning has been usually associated with thunderstorms. Some reports describe balls that eventually explode and leave behind a sulfur odor.

Officer José M. de los Santos drove toward the ball of light to investigate and woke up in the ditch a half-hour later with slight burns around his eyes. The windshield and one of the headlights of his 1970 Seat 1500 were smashed. The radio antenna of his police car was sharply bent backward, and his wristwatch and the clock on the dashboard were ticking 15 minutes slow. José always made sure on a daily basis that his wristwatch and the clock on the dashboard agreed with the very accurate clock at the police station. It was important for him to be accurate on his reports and for his own time keeping on the job. Something had made time dilate by 15 minutes. What could that possibly be? Would bolt lightning be able to slow down the passage of time?

Vanessa was very intrigued about the policeman story and felt consternation about her own experience. Little did she know that many years later her daughter Victoria would know more, through her work, about the possible source of the mysterious force field that hit and undulated parts of her car.

§ 8. *Is the technology for a warp drive or a hyperdrive real or science fiction?*

Let us imagine the science fiction journey of a fictitious very talented writer Victoria Quintus through the Aztec Kalakoayan (The Stargate in Nahuatl, the language of the Aztecs) into a fantastic realm where the advanced technology to travel through the vastness of our universe, and through the depth of time, exists. Let us first describe Victoria and her background. Victoria was born in beautiful Madrid, Spain, where she studied at a private catholic school for girls and graduated at the top of her class from the renowned Complutense University of Madrid where she studied physics, engineering, and business administration.

Victoria worked for various companies in the areas of engineering, design, and marketing communications, becoming a successful executive with no time to write or to draw, not doing all those things she liked, but as destiny would have it, her company was sold to a multi-national company. She started working in the area of communications and doing what she loved the most, writing books and telling science fiction stories for all ages. Finally! Fate guided her toward what she loves to do the most and where her great talent has blossomed.

Victoria has written one of the greatest youth adventure of recent years, The Aztec Kalakoayan (The Aztec Stargate), a series of fantastic stories that begin within the fantasy genre although the series ends up reaching science fiction in the last volume. She always has in her novels a fantastic ingredient, but she has also written adventures that are purely realistic. There are many characters, places, and phrases in the plot of her books that people think are made up but are actually real. Unbeknown to her readers, Victoria has been working as the lead project manager on one of the most technologically advanced projects for a private equity firm working for an international consortium of nations, thanks to her extensive background in physics and engineering.

Therefore, let us imagine our science fiction narrative with Victoria's ingredient that even though characters or settings can be made up, other parts of the story can be real.

A transmission was received by the Secretary-General of the United Nations from deep space, that requested a meeting with a leading scientist from earth regarding a very important exchange of information for our planet. The unknown sender refers to a highly classified international project on earth code-named "The Aztec Stargate" as the main reason for the meeting with the lead project manager and scientist, the very talented science fiction writer Victoria Quintus, whom they were able to identify through their advanced spying technology, for a specific date, location, and time on earth. The United Nations immediately contacted the project lead manager to arrange the meeting in total secrecy. A date and time was set for the desired location on earth for Victoria to meet with the designated extraterrestrial representative. The authorities would be monitoring the meeting from afar and snooping through Victoria's state-of-the-art smartphone.

As Victoria enters the Aztec Stargate, she instantly finds herself in another realm, it is a desert landscape full of sagebrush, the shrubs that are native to the North American West. She has landed in the designated desert region in northern Mexico, a few miles south of the U.S. border. Not far from the stargate, she sees what appears to be a glittering metallic disc, it looks like a spacecraft with an opening or door on its side. As she approaches the spacecraft curiously, she hears the buzzing sound of the stargate returning to its base, its departure location on earth. She turns and catches a glimpse of the disappearing stargate. The stargate is set to return after a specific period of time that is only known to Victoria.

As Victoria walks closer to the spacecraft, a woman steps out of the vehicle and greets her in an unknown incomprehensible language, but soon after adjusting her earbud, the woman speaks again, and Victoria could now understand her perfectly.

- The woman says to her in an electronic voice, "My name is Star and I have been sent to speak to you about our technology as requested by my superiors, since your stargate has a very limited range of operation and may be hazardous. My superiors have decided to provide you with our safe technology for space travel."
- After those words, Victoria stood there pensive, and then asked, why do you do this?
- Because we have researched your stargate and concluded that your stargate is unstable and emits hazardous gravitational radiation that interferes with our implosive hyper-space travel technology and the technology of other nearby civilizations in our galactic federation. It also affects the stability of your planetary core. Similar to the effect of massive thermonuclear weapons where an implosion creates an explosion.
- Victoria said carefully not to be ungrateful, "I appreciate your good intentions, but I would need to know more before I could proceed on my side to stop the most advanced transportation project that we presently have under development."

113

- Yes, I understand. I will start by describing the technology of our spacecraft to travel through hyper-space, and I will also record our conversation and provide you with a transcript.
- Let us walk around the vehicle as I explain the technology. An advanced spacecraft may need to use a combination of spatiotemporal drive systems to travel up to and beyond luminal speed. The primary spatiotemporal drive, or warp drive, would be the propulsion system to attain luminal speed, and a secondary spatiotemporal drive, or hyperdrive, to jump into hyper-space, or nil-time, for super luminal travel. At the start of the journey, the spacecraft would accelerate up to luminal speed, protected by several force fields for redundant safety, to prevent harm to the crew or any damage to the integrity of the ship's hull. Preceding luminal speed, the force fields may be lowered to allow for full dilation or contraction of mass and space-time. The full relativistic space-time-mass dilation would be necessary to initiate the jump of the spacecraft into hyper-space. From previous theoretical research, the mass and time dilate in a relativistic way as the spacecraft travels through space-time, then near the light barrier, the spacecraft jumps into hyper-space or nil-time, where the dilated mass condenses into rest mass, space contracts, and time condenses into nil-time. As the systems of mass in the spacecraft condense, the frequency of the wave of every particle, or macroscopic object, increases, and other wave properties change, during the hyper-space jump.
- Most of the travel time would be spent bringing the spacecraft up to luminal speed, and after a hyper-space jump, slowing it down to cruising subluminal speed at a safe distance from the final destination.
- Ask any question you would like about the spacecraft.
- Can you be more specific about the primary and secondary spatiotemporal drives? Asked Victoria.
- Yes, let us step inside the vehicle.

Both women entered the vehicle through the narrow plug door.

- I will answer your question. The primary spatiotemporal drive may be what you would call an Alcubierre's warp drive, according to information available on your internet, that would be able to modulate space by creating a distortion around it to hover, or in front of the spacecraft, or in any other direction of

114

propagation, with an Alpha gravitational field from a heavy poor metal reactor. The secondary spatiotemporal drive may be a bosonic field drive that would consist of multiple fiber optic coils around the geometry of the spacecraft that could work in tandem to balance, assist, or counteract, the unbalanced forces around the vehicle. The warp drive or the bosonic field drive may produce enough gravitational acceleration inside the spacecraft using wave guides and resonators and outside the outer hull to dampen, or cancel, inertia during rapid acceleration or deceleration. The crew of the spacecraft may use spacesuits that include a force field for redundant safety.

- The technology of the primary spatiotemporal drive may be regarded as an implosive technology. The heavy poor metal reactor continuously transmutes the masses and the gravitational fields of imploded atomic structures producing radiation and gravitational waves of the Alpha field to be emitted through the bottom of the spacecraft to create the necessary spatiotemporal distortion in the Beta field, or in free space, around the spacecraft to be emitted through a tuned tube, or stack at the top of the vehicle. As the Alpha field is emitted, it is guided through the tuned tube up to and over the shielded dome of the reactor as a gravitational shield and distributed through the wall waveguides throughout the spacecraft and through the floor of the cockpit to provide a stable gravitational field for the crew. The radiation and heat may be converted to current for electrical systems on board.

- What about the hull construction and the communication system? Asked Victoria.

- The hull construction and communication systems are also crucial technologies as I will explain to you next.

- The spacecraft would need to have a hull that is made of a conductive metal alloy, like silver or copper, and must be very resistant to environmental corrosion, rust, and oxidation, like nickel or gold. The spacecraft may be printed using an advanced molten metal printing technology, or an injection molding manufacturing process, for greater strength of the alloy material, reduced cost, and better uniformity without seams, defects, or welding. The spacecraft may use window panels that may act as transparent windows or sensors for the outside environment, to protect the crew of hazardous atmospheres or environments, and provide safe access to or egress from the vehicle and compartments.

- The outer hull of the spacecraft is to a large extent symmetrical with an aerodynamic design for travel in the atmosphere of celestial bodies. The force field around the geometry of the spacecraft creates a spatiotemporal displacement around the spacecraft that allows space-time and the atmosphere to flow around the contour of the vehicle to reduce inertia and aerodynamic drag. Hence, the crew only feels the gravitational acceleration of the Alpha field of the primary spatiotemporal drive but feels very little gravitational effect from the Beta field of the celestial body. The spatiotemporal wavelets would not expand or contract against the surface area of the vehicle or any system of mass related to the vehicle. The spacecraft would create its own gravitational effect. The gravitational waves around the spacecraft may affect the line of sight around the vehicle since images may be distorted as space-time curves. An observer looking straight up at the bottom of the spacecraft may only see what is above.

- As a spacecraft hovers in the atmosphere of a celestial body that has a magnetic field, like the earth, the primary spatiotemporal drive of the ship may generate a high voltage potential that may be visible as a bluish corona discharge on the bottom of the vehicle that undulates in the external and variable magnetic field. The spacecraft is hovering in space-time and gliding on the external magnetic field. From previous theoretical research, it is important to point out that space-time has an electromagnetic aspect to it, and vice versa.

- The hyper-space communication systems between spacecrafts may be achieved through the frequency modulation of a single tachyon as the tachyon moves through nil-time for the instantaneous transmission and reception of a stream of single tachyons. The tachyonic frequency modulation may encode information in as many single tachyons of a specific frequency, as necessary. The transmission may be started during the spatiotemporal modulation of a specific space-time-mass dilation process through hyper-space heterodyning between distant dimensional points to deliver a signal at much lower frequency. The tachyonic signal is decoded to the original message at the receiver. A tachyonic communication system may multiplex the simultaneous transmission of several messages along a single

116

tachyonic channel of communication between spacecrafts. Tachyon-communications may open a new medium of research and applications in deep hyper-space optical communications, single-tachyon laser ranging, as well as testing the fundamental principles of tachyonic physics in hyper-space for spacecrafts. Some of our computers work with a similar frequency modulation technology where the computer circuits and their elements can reconfigure, or rearrange themselves through resonant vibration and nano technology, as required by the frequency of a fundamental signal, for distinct stable functional operations.

– Can you expand on the secondary spatiotemporal drive? Asked Victoria.

– "Yes, I will give you more information." Star replied, as both women walked by the secondary spatiotemporal drive.

– The secondary spatiotemporal drive, or hyperdrive, engages for superluminal travel as the force fields of the spacecraft are lowered to enable the full effect of space-time-mass contraction or dilation to jump into hyper-space. At that moment, the dilated mass and dilated time contract, space extends in all directions about the geometry of the spacecraft, still exerting a gravitational effect on systems of mass, in a negligible period of time or nil-time. It begs the question; how would this process affect the health and well-being of the crew? As space-time-mass contracts or dilates, the spacecraft travels to a distant location instantly, to start deceleration from superluminal speed, which brings the vehicle back to a subluminal speed in the space-time at a safe distance from the final destination. Hyper-space travel may be visualized as the sliding of the system of mass, in this case, a spacecraft or objects of matter, through the spatiotemporal dimensions of the worldtubes of every traveling object, as the hyper-space medium remains static during a nil-time. Consequently, the crew of the spacecraft would feel no change in the gravitational effect of the Alpha field during the nil-time of hyper-space travel.

– The nearly timeless process of hyper-space travel needs to be extremely accurate to avoid traveling through time into the past instead of traveling through space to a distant location. The process of time travel is similar but may lead to alternate realities, or other challenging consequences or mishaps for the spacecraft and crew. The passage of time continues to be measured at the clock of the departure location, while the clock in nil-time is at tick-tock-zero, allowing the temporal distance between the location of departure and the spacecraft to pass rapidly into very long periods of time. To be lost in hyper-space is to be lost in time. It is possible that some civilizations in our universe may have been started by hyper-space travelers that have been misguided in time by their dead reckoning technology.

— During hyper-space travel, time and space are counterbalanced, the spatial force offsets the reciprocal divergent temporal force, which results in nil-time. Hence, to travel in time is in essence to travel through space to offset the divergence of space over time. It is a form of travel through static space and nil-time due to the spatiotemporal pressure differential between the departure point and the destination point in hyper-space. The clock on the spacecraft does not tick tock as the clock is moved by the pressure differential across the interdimensional interval at an indeterminate speed. It is the spatiotemporal flow about the contour of the spacecraft, or systems of mass, that provides the spatial displacement, while space is not diverging, or time is not passing. The on-board clock ticks at the departure point and tocks at the destination point. This is a form of interdimensional travel between two dimensional points, on separate unrelated light cones in the same universe through hyper-space. Thus, we may consider hyper-space as the space-time between unrelated light cones regardless of spatial separation. It is interesting to point out that this concept would change your current idea of causality between two distant objects in space-time.

After exiting the spacecraft, the women just stood by each other's side as they contemplated the outside features and contour of the vehicle.

— Is this spacecraft capable of time travel? Asked Victoria.
— Yes, it is. Let me explain.
— If the spacecraft were to travel to the future using the hyperdrive, then divergence of space and time may be counterbalanced (nil-time) inside the Alpha field domain, while outside the field domain, space diverges significantly more, and time passes. However, if the spacecraft were to be near the event horizon of a black hole, at a safe distance, it is possible for space to slightly diverge and for time to pass slowly, while space diverges, and

118

time passes noticeably more. As time passes around the spacecraft and the crew, all systems of mass travel to the future experiencing time dilation on the on-board clock compared to a reference clock on an inertial frame of reference at a significant distance from the event horizon. The future may be described as the fluid entanglement of objects and the evolution of the present events of those objects that persists in an observer's reality.

– The spatiotemporal modulation of the mass dilation effect by the hyperdrive, before or after jumping into hyper-space, allows the spacecraft to adjust the spatial divergence to a desired rate of flow. In a sense, the Alpha field modulates the flow the pressure differential of hyperspace about the spacecraft and the systems of mass. The on-board clock becomes relativistic according to the adjustment. If it were possible adjust the hyperdrive to halt the hyper-space flow, the spacecraft may be able to stay in nil-time for an indeterminate amount of time. Space-time is infinite and eternal, with infinite interspaces in the direction of the advanced or retarded wavefunctions. The spatial divergence or convergence is transcendental and fractal. This is the result of the infinite number of probable spatiotemporal waves and frequencies that form our reality. An infinite number of harmonics that add up to the fundamental wave of reality. It is all a matter of waves!

– I hope I have answered accurately your questions about our travel technology. We are also willing to assist you in developing and building a safe hyper-space transportation technology.

– Yes, you have. Thank you! but I am sure there will be more questions.

As Victoria stood there trying to internalize all the details that Star had given her, she could not help thinking of the words that she admired from the talented writer and theologian C.S. Lewis, "the future is something which everyone reaches at the rate of sixty minutes an hour, whatever he does, whoever he is. The past is frozen and no longer flows, and the present is all lit up with eternal rays. Hardships often prepare ordinary people for an extraordinary destiny."

Star nodded and smiled, knowing Victoria was pleased with her presentation and proposal. The two women agreed on a follow-up meeting to discuss further details about the technology transfer. After saying goodbye, Star entered the spacecraft and gradually lifted off toward her rendezvous with her mothership at a prearranged location in space near earth.

On her way to the stargate landing site, Victoria asked her back-up operatives if they had heard the conversation, and they replied "negative, it was an unintelligible language, and the signal was continuously braking up." Victoria answered, "Understood!" Then, she heard the familiar buzzing sound of the stargate materializing at a short walk from her position, at the same location where she had landed and said "My ride is here. I gotta go! See you all at the base."

As she walked to the stargate, her thoughts turned to the ulterior motive why the international consortium of nations had used the modified stargate to emit gravitational radiation. They had found a spacecraft, as well as a stargate, of unknown origin, in an archeological dig years ago near the Teotihuacan pyramids in Mexico, but the scientists involved in the research did not understand the advanced technology yet. They were able to power up the stargate to do short range hyper-space jumps, but not the spacecraft. The powers that be had proposed to bait the extraterrestrials with the stargate to motivate an encounter of the fifth kind. The strategy seemed to be working. The benefits for humanity would be tremendous. Victoria stepped through the stargate as planned, but neither the talented Victoria, nor the stargate, nor the extraterrestrials, were ever seen again. Fool me once, shame on thee, fool me twice, shame on me!

§ 9. Could energy be extracted from a theoretical white hole?

Researchers have recently confirmed the theory of the extraction of energy from a black hole. A Kerr-Newman black hole has eleven components of mass-energy, linear and angular momentum, position, and electric charge. However, the question remains if a similar theory could be use on a hypothetical white hole. Nothing could enter a white hole past the outer event ektropí, not even light. Hypothetical white holes can belch matter, particles, or energy, so white holes have hair. Thus, except for quantum fluctuations, stable hypothetical white holes may be completely described at any moment in time by the hairy theorem.

The hairy theorem states that all white hole solutions of the Einstein–Maxwell equations of anti-gravitation and electromagnetism in General Relativity cannot only be completely characterized by eleven externally observable classical components: mass-energy, linear and angular momentum, position, and electric charge, but also by other information that may come out from the interior of the white hole.

The Greek word Argos means idle. The Argos-Sphere is the outermost layer of its outer event ektropí or outer event deflection. In a hypothetical Kerr-Newman white hole there would also be an inner event ektropí. The boundaries may be regarded as mathematical surfaces, or as the boundaries of the surfaces of the physical fields acting within or without the Kerr-Newman white hole.

Hence, could it be technologically possible for an advanced civilization to extract a significant amount of energy from a hypothetical white hole if an object of mass or energy traveling at relativistic speed somehow changes the rotational energy? What would be the useful application of that technology?

Would a hypothetical Kerr-Newman white hole be only instantaneous rather than continuous or long-lasting? Would a white hole between a universe and a parallel universe be stable and long-lasting? Not long ago, both black holes and white holes were entirely theoretical, but now many actual astronomical objects are associated with black holes, and hypothetical white holes are thought not to be continuously observable, even though their effect may only be detected around the event itself. It is possible to ask the following rhetorical questions: What role would a black hole-white hole pair play between adjacent or kindred universes? In such scenario, a black hole may have a white hole at the other end that springs new universes into existence. A white hole in our universe may be the egress of a progenitor universe, a "Pachamama" in the Aymara or Quechua language of South America. Could a white hole be a spatiotemporal pressure regulator, a big bang event initiator, a mass-energy stabilizer, or a fast-spinning pulsar? Could gamma-ray bursts may also come from white holes and not only associated to supernovae? Would white holes from an adjacent universe explain why matter was favored over anti-matter in our early universe?

What if a white hole is a black hole that is moving backward in time on the advanced wave? Is that why we perceive a white hole the way we do and why we have not found any as we move forward in time? A black hole moves forward in time in the retarded wave. The black hole-white hole pair may be spatiotemporally concentric and may be considered a quantum superposition. Consequently, the black hole obeys the second law of thermodynamics in our universe in the retarded wave direction while a white hole obeys the reverse second law of thermodynamics in the advanced wave direction in the same universe. An interdimensional spatiotemporal bridge resembles a light cone for a black hole-white hole pair. A light cone or gateway between the present and the past or between kindred universes. The size of the throat of a black hole-white hole pair between present and past would be very infinitesimal at any given instant of time. The singularity would be shared by the black hole-white hole pair as the pair overlaps at any given spatiotemporal point in its geometry. The retarded singularity would consist of particles and the advanced singularity would consist of anti-particles. The black hole would exist toward the future and the white hole would exist toward the past of the same universe as implied in one of the solutions to the Einstein Field Equations of General Relativity.

Information from our universe that enters the black hole going forward in time in the retarded wave would not be lost forever, but the information would come right back to its source in the advanced wave of the same universe. Nothing would be taken from the universe but only reflected to its source. The entropy of the black hole would increase in the direction of the retarded wave and decrease in the direction of the advanced wave. There would not be a black hole information paradox, but instead an information rebound hypothesis. Information would not have to be stored on the boundary of a black hole. Black holes may still grow through the accretion of other black holes. This would be a solution to an important puzzle about black holes, that even though black holes appear to stay at a constant volume as viewed from outside of their outer event horizon, their interiors would not have to keep growing in volume essentially forever. The complexity of a black hole refers to a measure of the number of computations that would be needed to recover the initial quantum state when a black hole is formed. There would be no need for a growing complexity within the outer event horizon, the interior volume of the black hole would not have to grow bigger continuously as space extends toward the singularity.

Would breaking the light barrier of the retarded wave to travel in the advanced wave within a black hole-white hole pair, in a zig zag trajectory, result in the realization of a directional and traversable spatiotemporal bridge in our universe? Let us imagine an advanced technology that allows us to travel faster than light within the outer event horizon of a black hole, breaking the light barrier at a given spatiotemporal point of the retarded wave by traveling in the direction of the advanced wave within the outer event ektropí of the associated white hole.

It is possible to theorize that the hyperdrive of such a spaceship would allow a chrononaut to choose a point of exit in the past of his trajectory by traveling faster than light outside the outer event ektropí of the white hole reversing its direction toward the retarded wave to choose a specific temporal destination to exit in the past in the forward direction of time.

For the return trip in the forward direction of time, the chrononaut's spaceship may orbit the black hole in the past at a distance that would dilate time in the direction of the retarded wave while the time at a significant distance from the black hole is passing incredibly fast in the same direction until the spaceship exits at the specific temporal destination in the future.

White holes that are only microscopic may be very massive, a white hole the size of a grain of sand may weigh more than Callisto, one of Jupiter's moon, and the third-largest moon in the solar system. It is possible to theorize that if a black hole in our universe meets a white hole from another universe, the result might be a single larger black hole-white hole pair, or an interdimensional bridge, across multiple universes.

Conversely to a Kerr-Newman black hole, it can be predicted that an object of mass, in a Kerr-Newman white hole, would acquire positive energy. The object would have to be in the argos-sphere traveling at a relativistic speed and would have to be divided in two. In this way, one half would have lower mass-energy while it is deflected away from the white hole, and given the recoil action, the other would be spiraling closer to the outer event ektropí of the white hole with greater positive mass-energy.

The insertion of energy occurs in the rotational energy of the white hole outside the outer event ektropí in the Kerr-Newman spatiotemporal region called the argos-sphere. All objects in the argos-sphere become dragged by a rotating space-time. Thus, the spiraling half would transfer energy that would be inserted into the rotational energy of the white hole's own argos-sphere.
Even though momentum is conserved the effect is that more energy can be inserted into the white hole than was originally provided, the difference being absorbed by the white hole itself.

The insertion process, or Hawking process, results in a small increase in the angular momentum of the white hole, which corresponds to an energy transfer from the object of mass. The energy absorbed is converted into a gain of momentum.

The maximum amount of energy transfer predicted for a single particle via an efficient process would be greater than twenty percent in the case of a charged rotating Kerr-Newman white hole. According to the Heisenberg's uncertainty principle of quantum mechanics, a rotating Kerr-Newman white hole would dismantle particle pairs and absorb or emit particles from within, which could be emitted to its argos-sphere.

A white hole is a radiation source. *It is possible to theorize a pair of virtual particles that emerge within the outer event ektropí of the white hole where one of the particles spirals closer to the white hole singularity while the other is deflected outside of the white hole as Penrose radiation.*

If such a process continued, it would be possible to predict that the white hole may lose its mass gradually through particle-anti particle annihilation or particle emission, increasing its volume, decreasing its rotation, while its associated black hole would discharge more mass from its singularity to the center of the white hole, decreasing the volume of the black hole, increasing its rotation, which would continue until the black hole-white hole pair contracted and dissipated.

However, as long as the black hole fed on more matter or energy, the white hole would sustain and balance the black hole-white hole dipole, through the principles of conservation of energy and conservation of charge, without complete dissipation of the pair.

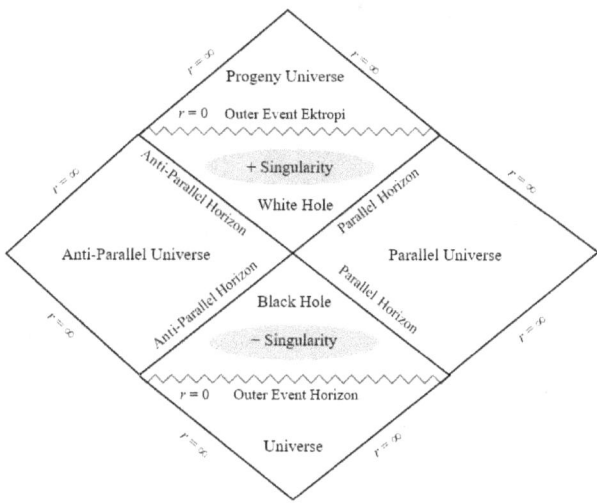

Figure 3. An Interdimensional Spatiotemporal Bridge Diagram

Consequently, it is possible to predict a Penrose radiation to describe the emission of particles or objects of mass, or pure energy, from a Kerr-Newman white hole to the argos-sphere or to space. Therefore, white holes have a certain temperature, and an amount of entropy that is reciprocal to the surface area of the outer event ektropí. Thus, a black hole tends to increase its surface area as its entropy increases through the accretion of matter and energy toward its singularity, while a white hole tends to increase its surface area as its entropy decreases through the emission of matter and energy from its singularity.

$$S = \frac{1}{\pi R^2} = \frac{4}{A} \tag{9.1}$$

Where S is the entropy of the white hole, A is the area of the outer event ektropí, and R is the radius of the white hole.

Furthermore, every rotating white hole would lose energy in the form of anti-gravitational waves.

As the anti-gravitational waves travel through the plenum of matter they would extend the spatial dimension in the direction of propagation while a gravitational wave would contract it. Both are longitudinal waves that travel from their source by extension or contraction of the spatiotemporal divergence or convergence of the medium. As an anti-gravitational wave passes through an object of mass, it would extend the body of the object in the direction of propagation while it passes through. Anti-gravity is not the displacement of a gravitational wave by a hundred and eighty degrees as it would be for an Alpha gravitational field in a planetary gravitational field, or Beta gravitational field. A more appropriate name for the Alpha gravitational field would be a counter gravitational field.

Hence, the extending of the spatial dimension of anti-gravitational waves begs the rhetorical question: if a black hole can dilate time when space contracts, so that time slows down, can a hypothetical white hole contract and speed up time when space is extended?

Aside, the inner edge of a black hole is calculated from the measurement of the disk temperature and luminosity to infer the rotation of the black hole. According to researchers, a massive black hole known as ASASSN-14li rotates at least at half the speed of light; it completes one rotation in about two minutes.

Let us propose a Penrose-Zeldovich experiment to test the hypothesis of our theory by using spiraling waves of light that strike the surface of a lightweight Vantablack disk. Vantablack, or vertically aligned nanotube arrays that are black, is one of the darkest materials known, absorbing up to 99.965% of the perpendicular visible light striking the material. It is composed of a bundle of vertical tubes grown on a substrate.

When light strikes the material, it is continually absorbed and dissipated through the tubes as heat. The material has high thermal stability and high resistance to mechanical vibration. As the spiraling light strikes the rotating disk, at the right rotational speed, the momentum impacted by the spiraling light against the direction of rotation, would be absorbed by the disk and it would lower its rotational speed slightly, the amplitude would be attenuated, and the frequency of the light would be blue shifted (shorter wavelength).

126

The disk would absorb the energy due to the attribute of the doppler effect in the incident spiraling light. This effect would be similar to the effect of a hypothetical object positioned in the ergosphere of a black hole that splits into two halves, with one half absorbed by the black hole and the other half ejected away from the black hole with greater energy while the black hole's rotation slows down.

The experiment may also be performed with sound waves, a source of waves with a much lower frequency than visible light with a set of rotating miniature speakers and a rotating foam disk with very sensitive microphones behind the disk connected to a multitrack recorder. The rotating sound waves would be against the direction of rotation for maximum effect. The microphone would receive the rotating sound waves that are changed by the doppler effect as the foam disk slows down slightly.

Let us propose a similar experiment for a white hole, where spiraling microwave rays strike the metal reflecting surface, of a rotating disk. A few microns of silver or copper is enough to give a good reflection of microwaves. In which case, at the right rotational speed, the momentum of the spiraling microwave rays in the direction of rotation, would be reflected by the metal surface, raising the rotational speed of the disk, while the amplitude would be amplified, and the frequency of the microwave rays would be red shifted (longer wavelength). This effect is similar to the effect on spiraling microwave rays deflected from a hypothetical white hole with greater energy than the incident spiraling microwave rays. This experiment may also be performed with sound waves, with the disk of rotating speakers rotating in the same rotational direction as the receiving foam disk, as the foam disk speeds up slightly with greater energy at the right rotational speed.

Theoretically, as a spatiotemporal volume of wavelets converges through the event horizon of a Kerr-Newman black hole, the spatial radial length toward the singularity extends (spaghettification), while the perpendicular spatial lengths of width and depth associated with the spatiotemporal volume of wavelets contract toward the singularity due to the geometry of the black hole and the passage of time, or the convergence of space. It is possible to theorize that as spatiotemporal volume extends through and around the theoretical positive ring singularity of a Kerr-Newman black hole into the throat of the black hole-white hole pair, or interdimensional spatiotemporal bridge, the spatiotemporal pressure decreases, increasing the velocity of the spatiotemporal divergence rate through the interdimensional throat. As the lower pressure spatiotemporal volume extends out of the throat into the Kerr-Newman white hole, it expands through and around the theoretical negative ring singularity with even less spatiotemporal pressure.

The spatiotemporal divergence of lower pressure acts on and around the negative ring singularity creating a spatiotemporal region below it that may maintain the negative ring singularity within the white hole. This retaining effect keeps the negative ring singularity inside the white hole while the spatiotemporal pressure outside the outer event ektropí is much higher. As the black hole positive ring singularity and volume diminish causing the spatiotemporal pressure through the throat and inside the white hole to increase, the resulting pressure, and lesser positive charges in the black hole, would cause the remaining material of the negative ring singularity in the white hole to approach the outer event ektropí for eventual emission and dissipation.

As matter is accreted into the Kerr-Newman black hole, the mass condenses under enormous pressure into a ring singularity. It is theorized that the very dense and fluid cloud of electrons of the nuclei of the heavy elements in the ring singularity are displaced by the extending spatial length through the interdimensional throat into the realm of the associated white hole as an extended negative ring singularity while the heavy nuclei, in the realm of the black hole, form a positive ring singularity. Electrons could establish their bidirectional orbital trajectories based on the bidirectionality of spatiotemporal divergence, or convergence, between any two point within the wave medium.

Consequently, it is possible the charged ring singularities form an electromagnetic dipole, in which case, the ring singularity in the white hole would be retained by lower pressure as well as an electromagnetic force of attraction from the ring singularity of the black hole, extending the life cycle of the white hole proportionally to the life cycle process of the black hole. When spatiotemporal flow is convergent or divergent, the spatiotemporal density varies with its pressure. Convergent or divergent flows are luminal speed flows.

The Bernoulli equation can be adapted to flows that are convergent or divergent. However, the assumption that the shear forces due to viscosity are negligible (inviscid) remains in the convergent or divergent versions of the equation. The convergent and divergent effects depend on the speed of light in the spatiotemporal medium and the reciprocal relationship between space and time.

The Penrose-Venturi effect is the increase or decrease of static spatiotemporal pressure that results when a spatiotemporal volume of wavelets converges or diverges across two points of the throat of an interdimensional spatiotemporal bridge between a Kerr-Newman white hole and its associated black hole without any difference in potential, torsion, or electromagnetic pressures across the interdimensional bridge.

$$B_{\mu\nu} - W_{\mu\nu} = -\frac{G\rho}{2c^4}\left(c_{WH}^2 - c_{BH}^2\right) = \frac{G\rho}{2c^4}\left(c_{BH}^2 - c_{WH}^2\right) \tag{9.2}$$

where "$B_{\mu\nu}$" is the spatiotemporal curvature of the black hole, "$W_{\mu\nu}$" is the spatiotemporal curvature of the white hole, "ρ" is the mass density (kg/m³) of the ring singularity and "c" is the speed of light for a spatiotemporal volume of wavelets.

The above equation shows that, in the event of a curvature change or a change of spatiotemporal pressure, the speed of light would be determined by the reciprocal relationship between space and time, and vice versa. The Euler-Bernoulli spatiotemporal principle states that a change in the spatiotemporal divergence rate or convergence rate occurs simultaneously with a change in static spatiotemporal curvature or spatiotemporal pressure.

$$\Sigma = A_{BH}\sqrt{\frac{16\pi}{\kappa\rho} \cdot \frac{\left(B_{\mu\nu} - W_{\mu\nu}\right)}{\left(\dfrac{A_{BH}}{A_{WH}}\right)^2 - 1}} = -A_{WH}\sqrt{\frac{16\pi}{\kappa\rho} \cdot \frac{\left(B_{\mu\nu} - W_{\mu\nu}\right)}{1 - \left(\dfrac{A_{BH}}{A_{WH}}\right)^2}} \tag{9.3}$$

whereas "Σ" is the volumetric flow rate (m³/s), "κ" is Einstein's constant, "A" is the cross-section (m²) of the throat on each side of the bridge, and the other variables were described previously.

Consequently, assuming the geometry does not change, we can theorize the following effects: the higher the spatiotemporal pressure differential across the interdimensional throat, the higher the bidirectional velocity of the spatiotemporal divergence or convergence rate through the interdimensional throat, and the lower the density of the positive ring singularity in the black hole or the lower the spatiotemporal pressure differential across the interdimensional throat, the lower the bidirectional velocity of the spatiotemporal divergence or convergence rate through the interdimensional throat. Notwithstanding, the directional flow of mass-energy into the black hole or out of the white hole would continue.

Black Hole

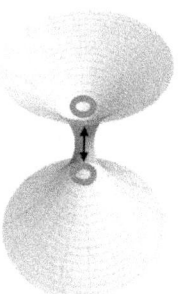

White Hole

Figure 4. Kerr-Newman Black Hole – White Hole Pair Illustration.

The Schwarzschild metric for a black hole – white hole pair may be expressed as

$$ds^2 = -e^{\mp 2\frac{gr}{c^2}}dt^2 + e^{\pm 2\frac{gr}{c^2}}dr^2 + r^2 d\Omega^2 \qquad (9.4)$$

Where $2gr^2/c^2r \equiv 2gr/c^2$, $e^{-2\frac{gr^2}{c^2 r}} \approx 1 - \frac{gr^2}{c^2 r} + ...$, and

$\left[e^{-2\frac{gr}{c^2}} \right]^{-1} = e^{+2\frac{gr}{c^2}}$ for the Schwarzschild metric.

The following equation is an adaptation of Euler's Bernoulli equation for spatiotemporal pressure:

$$P_{Static} + P_{Dynamic} + P_{Potential} + P_{Torsional} + P_{Electromagnetic} = P_{Total\ Spatiotemporal} \quad (9.5)$$

Torsion may be produced by frame dragging or swirl inside the black hole or its associated white hole, where space-time churns around a massive, rotating ring singularity. The phenomenon known as frame dragging is also known as the Lense-Thirring effect.

Thus, let us express the Euler-Bernoulli Spatiotemporal Principle with Torsion and Electromagnetism, for the spatiotemporal region of pressure across the throat of the interdimensional spatiotemporal bridge inside the black hole-white hole pair during its theorized life cycle.

For a convergent or divergent spatiotemporal flow, we obtain

$$\frac{c^4 B_{\mu\nu}}{G} + \frac{1}{2}\rho c_{BH}^2 + \rho g d_{BH} + \nabla_b L_{\mu\nu}^b + \frac{F_{\mu\nu}^+}{A_{BH}} \quad (9.6)$$

$$\frac{c^4 W_{\mu\nu}}{G} - \frac{1}{2}\rho c_{WH}^2 - \rho g d_{WH} - \nabla_w L_{\mu\nu}^w - \frac{F_{\mu\nu}^-}{A_{WH}} = \rho c^2$$

$$P_{BH} - P_{WH} = \rho c^2 \quad (9.7)$$

Where "c" is the constant speed of light, "g" is the gravitational or anti-gravitational acceleration across a gravitational or anti-gravitational potential, "L" is a force of torsion, "A" is a spatiotemporal area, "P" is pressure, "F" is an electromagnetic force, and "G" is the universal gravitational constant.

Torsion is pressure, or the twisting of a spatiotemporal volume due to a force of torque (N·m). Torsion may be expressed in Pascal units (Pa), or in Newtons/m^2.

Hence, when the black hole pressure is converging or diverging, the corresponding expression for the white hole pressure is diverging or converging, depending on the theorized life cycle of the black hole-white hole pair, and the principle of conservation of energy. Pressure is equivalent to energy density.

A particle of mass would follow a streamline trajectory, as a curve that at all points is tangent to the velocity vector of the particle, through the interdimensional throat. The mass flow rate would be equal to the product of the spatiotemporal flow density, the cross-sectional area, and the spatiotemporal velocity.

In terms of energy across the interdimensional throat for convergent or divergent spatiotemporal flow, we obtain

$$\frac{c^4 B_{\mu\nu}}{G\rho} + \frac{1}{2}c_{BH}^2 + gd_{BH} + \frac{\nabla_b L_{\mu\nu}^b}{\rho} + \frac{F_{\mu\nu}^+}{\rho A_{BH}} \tag{9.8}$$

$$-\frac{c^4 W_{\mu\nu}}{G\rho} - \frac{1}{2}c_{WH}^2 - gd_{WH} - \frac{\nabla_w L_{\mu\nu}^w}{\rho} - \frac{F_{\mu\nu}^-}{\rho A_{WH}} = c^2$$

In terms of a pressure head spatiotemporal distance across the interdimensional throat, we have

$$\frac{c^4 B_{\mu\nu}}{G\rho g} + \frac{1}{2}\frac{c_{BH}^2}{g} + d_{BH} + \frac{\nabla_b L_{\mu\nu}^b}{\rho g} + \frac{F_{\mu\nu}^+}{\rho g A_{BH}} \tag{9.9}$$

$$-\frac{c^4 W_{\mu\nu}}{G\rho g} - \frac{1}{2}\frac{c_{WH}^2}{g} - d_{WH} - \frac{\nabla_w L_{\mu\nu}^w}{\rho g} - \frac{F_{\mu\nu}^-}{\rho g A_{WH}} = ct$$

It was the eminent Leonhardt Euler who derived Bernoulli's equation in its usual form in 1752. Even though, the brilliant mathematician and physicist Daniel Bernoulli deduced that the speed of a fluid flow increases when the pressure decreases, for adiabatic and reversible processes that have the same entropy, when irreversible processes, like turbulence, and non-adiabatic processes, like heat radiation, are negligible.

Aside, all liquids are fluids but not all fluids are liquids. Liquids are the incompressible fluids; their density does not change significantly with pressure. Fluids on the other hand describe a substance that can flow as a result of a pressure differential between two points. Gases are also fluids and compressible. Space-time is neither a liquid nor a fluid. Some physicists may regard it as a superfluid.

Every point in a steadily diverging, or converging, and rotating spatiotemporal volume through the interdimensional throat, regardless of the spatiotemporal volumetric flow rate at that point, has its own unique static, dynamic, potential, torsional, and electromagnetic spatiotemporal pressures. In the absence of any other pressure, their sum may be defined as the total spatiotemporal pressure which may be regarded as a constant along the interdimensional throat.

White holes decrease entropy, which is a fundamental piece of evidence against them. In our universe, we obey the laws of thermodynamics. And so far, no confirmed violations have been observed in our universe, or widely demonstrated or duplicated experimentally, although we often hear claims of perpetual motion machines and unusual events. Could an alternate universe obey a different law for entropy? Are the laws of nature unique to a universe? Could a spaceship that came out of an alternate universe have a warp drive that works in violation of the first law or second law of thermodynamics of our universe?

A Kerr-Newman black hole has charge, so it is possible to hypothesize that a black hole and its associated white hole have opposite charges. Thus, let us hypothesize that a black hole-white hole pair is a dipole, with the black hole having a positive charge and the white hole having a negative charge. The positive charge of the black hole has an outward electromagnetic field, and the negative charge of the white hole has an inward electromagnetic field. Even though, when the direction of the spatiotemporal contraction of the black hole and the corresponding direction of the spatiotemporal expansion of the white hole disagree with the electromagnetic field directions, the spatiotemporal wavelets travel bidirectional between any two points inside the black hole or the white hole. Expansion or contraction are the resultant divergence, or convergence, of the interference of the spatiotemporal waves within each medium.

It is possible to extract energy from the rotating disk of material surrounding a black hole, and in fact when a small black hole absorbs a stellar mass, at least half of the rest-mass energy of the infalling material is expected to be radiated away before what is left enters the outer event horizon of the black hole, which is more, by orders of magnitude, than the energy produced by the most efficient nuclear fusion known, the conversion of hydrogen into helium. Furthermore, the incorporeal transmission of energy between celestial bodies results in rotation.

Let us imagine the science fiction journey of Captain David Quintus, in command of the UNERS Cisne, in a journey from earth to one of Jupiter's moons. A journey that ends where everything begins between universes!

Galileo VII was the latest spacecraft to examine Jupiter and its moons for an extended period of time. It was launched on an Ariane 5 rocket in 2056, got some speed boosts by swinging past Earth twice and Venus once, then arrived at Jupiter at last in 2062.

The Galileo VII probe was returning home from the Jovian mission and got pushed slightly off-course by the anti-gravitation of a white hole. It detected the white hole by chance on its return to earth after scanning Callisto. The white hole was located approximately four million miles (approximately six and a half million kilometers) from Callisto, one of Jupiter's moons. The UNS Cisne had been built by the UN Navy to be part of the United Nations Space Force Fleet, but since the discovery of the white hole near Callisto, it had been re-assigned to the white hole exploration and research mission called "Alba Foraminis" and the spaceship was renamed UNERS Cisne.

The UNERS Cisne is beautiful, it is an intersolar crystal cathedral or palace. The gothic look and feel of the UNERS Cisne is astonishing. The UNERS Cisne was modified for an experiment to extract energy from a white hole via rotating microwave rays that would hit the rotating white hole and ricochet at greater energy due to the Zeldovich effect to be picked by receiving antennae on a battery drone spaceship called UNBDS Alba, where a large-scale battery system would be located. The white hole was viewed as a potential free energy depot for the advancement of the United Nations Space Force where all nations would have an equal share of the space domain for the advancement of humanity. The government of the European Union had invested heavily into the United Nations Space Force with a leadership consensus free of political dogma, for the development of space technologies and the global defense of the earth and the moon from external threats.

The entire expedition was financed by the Columbus foundation for space exploration and research from Spain's southern spaceport at the Isla de Saltes in Palos de la Frontera with technical assistance from the UN and the EU. Palos de la Frontera is a town and municipality located in the southwestern Spanish province of Huelva, in the autonomous community of Andalusia.

Dr. Quintus was charismatic and auspicious. He was going to travel where no man has dared to go, to the outside of a white hole. His friends told him that he had to be nuts, it was impossible to survive near a white hole, because there would be a lot of radiation coming out of the white hole. But they knew that the word 'impossible' was not found in Dr. Quintus' dictionary.

His mother Victoria had disappeared in a very strange set of circumstances that were recently declassified by the United Nations. Once Victoria told David that "Every great journey begins with a first step. Every great idea begins with a visionary. Never forget, that you have within you, the creativity, the determination, and the foresight, to reach for the plenitude of galaxies to change the universe." Those motivating words made a lasting impression on the young boy.

Captain David Quintus had entered the United Nations General Space Academy at Marín, Pontevedra, in north-western Spain, graduating summa cum laude in 2045 with a PhD degree in physics. Subsequently, he entered the UN Navy's space propulsion program and after completing his propulsion training and spaceship school, Dr. Quintus joined the United Nations Space Force in 2049. After a very successful career in space exploration and research throughout the solar system he was a recipient of numerous space force awards which include the UN Legion of Merit, UN Meritorious Service Medal, UN Space Force Achievement Medal, and was a recipient of the UN Command in Space Badge.

On a beautiful and sunny summer day, a UN Ariane 5 launch vehicle had taken a Hermes spaceplane transporting Captain Quintus and some of his officers to earth orbit to proceed to a moon orbit where the UNERS Cisne awaited them. A few hours after arrival, the captain called for a highly classified mission briefing.

During the mission briefing, Captain Quintus, planned to explain the details of the experiment and the research mission as well as answer any questions from the crew of the UNERS Cisne.

- Some of you may be wondering why this team has been assembled. Our commanding officer Captain David Quintus will discuss our highly classified mission and answer any questions you may have. Said the executive officer.
- Captain Quintus walked on the dais and addressed the crew.
- Please feel free to ask any questions as I discuss details of our mission.
- Some weeks ago, one of our probes on its return path from Callisto discovered a white hole by accident. Through thermal imaging we generated a three dimensional image of the outer event ektropí of the white hole. Experts cannot agree on who put it there and when. However, contrary to our expectations, it is stable.
- The UNERS Cisne has been fitted with the equipment to test the effect of spiraling microwave rays deflected from a white hole with greater energy than the incident spiraling microwave rays at a safe distance from the white hole.
- Captain, what about radiation coming out of the white hole? Asked one of the medical officers.
- If you were to approach a white hole in a spacecraft, you could be irradiated by an enormous amount of energy, which would most likely destroy your ship. Even if your spaceship could withstand gamma rays, light itself would start slowing you down like the aerodynamic drag slowing down an airplane in the atmosphere of the earth, unless the spaceship had a spatiotemporal shield from an Alpha gravitational field. Even though our spaceship will be at a safe distance from the white hole, we have two Alpha gravitational field reactors on the UNERS Cisne that also provide us with an earth-like gravitational field on the spaceship.
- Sir, would spatiotemporal curvature be an issue? Asked a crew member.
- Even if the spaceship were built to be unaffected by the energy emission, space-time would be strangely warped around a white hole; approaching a white hole would be like going uphill. The

136

acceleration required would get higher and higher while the spaceship would move less and less. There is not enough energy in the universe to get you inside. How could energy inside a white hole seemingly come from nowhere other than space-time itself? These are reasons why their existence in our universe were doubtful. Our spaceship will be at a safe distance from harmful effects due to spatiotemporal curvature or radiation.

- To observers on a spaceship, a white hole looks similar to a black hole. It has the same attributes as a black hole, but if the observers kept watching they may witness a belch event, the moment when matter or energy comes out of the white hole, and they would say "Wow, this is really a white hole!"
- I am of the school of thought that every charged rotating black hole may have an associated white hole. The interior of a white hole in our universe is outside the causality of our universe. No outside event will affect the inside of the white hole.
- Are white holes even supposed to exist in our universe? Asked a crew member.
- Some researchers used to think that white holes cannot exist, because white holes are a misunderstanding of an early, incorrect analysis of an oversimplified theory of the nature of black holes. Once the simplifying assumptions were removed, so that a black hole could be modeled, all that remained was the entrance to the black hole on one side of the multi-dimensional space-time diagram, and the singularity in the center. White holes decrease entropy, which is a fundamental piece of evidence against their existence at this time. In our universe, we obey the laws of thermodynamics. And so far, no confirmed violations have been observed.
- What about people or devices that may exist in an alternate universe, do they obey different laws? Asked a crew member.
- Well, one reason people think that a white hole should exist is the mystery of what happens to all the spiraling matter around a black hole that falls in, beyond the outer event horizon, plummeting toward the singularity, into what nobody knows. Then, what?
- The Big Bang may be regarded as a white hole. But white holes are not infinite. A white hole may have a particular kind of singularity: a hypothetical naked singularity. There was no hard evidence proving that white holes existed until now, but in our vast and complex universe, there's space even for them.

- A white hole's outer event ektropí is a boundary of no admission, a perfect cloak. Spaceships from an adjacent universe may hide behind the boundary since they would be unseen and undetected. A perfect place to hide an ominous weapon or a base of operation. However, any electromagnetic signal that would be emitted by equipment within the white hole could be picked up outside of the white hole, that means you can never send any information into the white hole, it only comes out of it.
- Captain, I think your experts have the evidence now. It exists! Said one of the control room officers. "The Einstein field equations struck physics more than a century ago, like a hurricane through Homestead, and physicists and researchers are still sorting through the debris. A white hole is a possible solution to the Einstein Field Equations of General Relativity. A white hole is the time-reversal of a black hole."
- Yes, one of the solutions to the Einstein Field Equations of General Relativity has a black hole region in the future and a white hole region in its past. Thus, white holes appear in the theory of eternal black holes. A white hole is the opposite of a black hole. A white hole is a region of space-time that cannot be entered from the outside, even though matter, light, energy, and information, can escape from the white hole. A white hole, like a black hole, has properties like charge, mass, and angular momentum.
- Captain, what will be the anti-gravitational acceleration from the white hole? Asked a communications officer.
- Please remember that we are going into uncharted territory, so, according to researchers, the anti-gravitational acceleration of a white hole is expected to be $g = +rc^2/a^2$, where "r" is the distance from the white hole, and the radius of the white hole with a naked singularity is expected to be given by $a = 2GM/c^2$. A white hole with a naked singularity has repulsive force. Therefore, it radiates anti-gravitational waves. In other words, the white hole rotates with an expected angular frequency of c/a. The expression c/a is also called the Hubble constant. The energy of the rotating white hole is expected to be obtained at the angular frequency of emission. By the law of conservation of angular momentum, as the radius of a rotating object decreases, its rotational speed increases.
- Supermassive black holes are theoretically predicted and located at the center of galaxies to form galaxies. Is it possible that those

supermassive objects also exist as white holes during the big bang of a universe or a small bang of a galaxy? Could a white hole collapse from a tiny perturbation? Asked a crew member.

- Yes, I think it is possible that supermassive black holes have their associated white holes. However, a tiny perturbation is unlikely to cause the collapse of a hypothetical white hole into a black hole since nothing can enter a white hole from its surrounding space and the cause and effect of events inside and outside the white hole are mutually exclusive. As white hole expand gradually even at the speed of light, they are not explosive, but rather slow compared to the vast distances of a potential universe. Since a white hole can either expand or contract, like a black hole, it may produce or consume energy in that process, or in the Penrose process.
- All of you have extensive experience in space in your field that is why you have been chosen for this mission. We think you are the best possible crew we have for this important mission.
- So, I want to reassure you that we have planned our mission carefully to be as safe as possible within the constraints of the mission.
- Are there any other questions? After waiting during several minutes of silence, Captain Quintus said "if there are no other questions this briefing has ended. Thank you all."
- After leaving the meeting room when they were by themselves, the executive officer asked Captain Quintus, "how do you think it went, Captain? Some people were concerned about safety even though I know we are prepared."
- You are right about that, but the train has left the station, and I think I speak for everyone in our crew. This is worth the risk. Thanks XO, for your help and support.

The Ship's log last entry: Callisto 0800 The UNERS Cisne has arrived at its destination for mission Alba Foraminis.

As the UNERS Cisne was doing its countdown to unleash the powerful microwave beam at the white hole, some of the officers observed from the command deck that there was a flash of light out of the white hole where an unobserved silver disc flew out and made a banking turn to the right through the intended trajectory of the microwave beam, then it glided away from the white hole. The flash of light occurred as the disc automatically powered down its spatiotemporal hyperdrive as the disc glided to engage its luminal warp drive. Since the crew was at a safe distance and totally dazzled by the yellowish-bluish light of the halo and the yellowish-bluish fuzzy contour of the white hole, the crew was unfazed by the event, even though an object was tracked on the ship's radar which was taken to be a belch from the white hole. A few minutes later, the powerful microwave beam was activated onto the white hole and adjusted to ricochet exactly on the enormous antennae on the UNBDS Alba. The amount of energy in such a concentrated microwave beam was incredibly powerful. The experiment was working as planned, the energy extracted exceeded their expectation and it was increasing, and radiation from the white hole was within the expected safe range.

The crew was elated as they looked at the effect of the ricochet on the outer event ektropí of the white hole. Suddenly, there was another flash out of the white hole where an identical disc flew out in the same trajectory but as the second disc made a banking turn to the right, it was cut in half through its center by the powerful microwave beam. The crew of the UNERS Cisne was observing the beam trajectory attentively and were shocked to see a very bright flash of light as one half of the disc was deflected to the right at a lower speed beneath the halo and the other half climbed in a spiral rapidly and irregularly toward the halo of the white hole at a higher speed through the argos-sphere, according to the ship's radar, before vanishing. The explosion and the slicing of the disc happened incredibly fast, and it took a while for the crew to power down the beam and to realize what had happened.

Soon after, the control room received an unexpected communication from an unknown spaceship nearby.

- Earth spaceship, this is the pilot of the first extraterrestrial spaceship that came out of the white hole.
- This is control at the UNERS Cisne, go ahead.
- I request permission to board your ship. I am from earth.
- There was no immediate reply coming from the UNERS Cisne but after a few minutes they responded. "Understood. Please approach the ship by airlock 3 on the right side or stern. An officer will meet you at the airlock, over."
- Will do. Thank you.

The disc approached slowly and extended its brow to the UNERS Cisne's quarterdeck, several spaceship artillery pieces were trained on the disc. The unexpected astronaut exited the disc in an unusual spacesuit and unarmed, holding onto the railings of the brow and came to a halt at the end of the brow turned toward the stern of the spaceship where a well-armed officer of the deck in a spacesuit was waiting, then asked for permission to come aboard and rendered the officer a hand salute. The officer granted permission by saying very well and returned the hand salute. The daring officer on the quarterdeck was Captain David Quintus. As the astronaut approached him, David could not believe his eyes as he could now see the astronaut's face; the astronaut was none other than his mother Victoria whom he had not seen since he was a very young boy.

The airlock opened slowly, and David led the way in for Victoria into the interior of the large quarantine area of the ship. All kinds of questions and thoughts were running through David's mind, but he was overwhelmed by the emotions of seeing his mother again after all those years. Nonetheless, he felt happy at the same time because he knew they were going to have plenty of time to catch up and be a family again. He thought, this is going to be interesting. He recalled his mother saying, "when you're true to who you are, amazing things happen."

§ 10. Could cosmic rays come from the earth?

Cosmic rays are thought to emanate from the throats of spinning black holes or from the explosion of supernova, not from the matter that makes up a planet. Cosmic rays are expected to interact with the atmosphere of a planet, like the earth, as they hit the atmosphere or reflect off the ice sheet of Antarctica, which would produce vertically polarized pulses of radiation. Cosmic rays consist of high-energy protons and atomic nuclei which have very large cross sections compared to other particles such as neutrinos which can move through the atmosphere of a planet as if the planet were not there.

Is it possible that the upward-pointing cosmic rays detected as coming from the ice sheet of Antarctica are not passing through the earth but instead are coming through and out of a spatiotemporal bridge? Could these spatiotemporal bridges occur naturally or are they created artificially through advanced technology?

As a spatiotemporal bridge is opened near a source of highly energetic particles, those particles and radiation that are traveling parallel to the opening portal of the bridge may have the possibility to travel across to the other side which may be light years away in the destination universe. If such scenario were possible, then the cosmic rays from some energetic places in the universe would be entering the opening of a spatiotemporal bridge at a remote location in the universe at close to the speed of light and exiting upward through the opposite end of the bridge above the continent of Antarctica. It is highly improbable, some researchers may even say impossible, that these cosmic rays, as we presently understand them, traveled through deep space and passed through planet earth and emerged on the other side without any interaction with the atmosphere, the matter of the earth, or the ice sheet of Antarctica. According to the current standard model, known particles would not be able to propagate all the way through the atmosphere and matter of the earth at these very high energies, steep angles of trajectory, and then exit with horizontal planes of polarization.

Chapter 7

Quotes by Categories

§ 1. Physics

- Nothing happens in contradiction to the laws of nature. Only in contradiction to our understanding of those laws.

- Everything in our universe was somewhere at once but then space-time distanced everything through expansion. The speed of time in our reality is very slow with respect to the vast spatial distances of our universe. The speed of time is the speed of causality.

- If the universe were not expanding, but some things were getting smaller at the speed of light or accelerating, would everything look like it was expanding from our perspective on earth?

- Space-time is a cosmic architect. Space-time is what the universe really was, is, and will be. Space-time is all there is.

- The wavefunction is spatiotemporal and space-time is all probability. Energy and mass distort space-time and that distortion to probability is spatiotemporal curvature. Probability times a force equals energy density. The rest is physics!

- If people find fault with Quantum Mechanics, let them be sure they understand it.

- If particles were self-aware in Quantum Mechanics, a nucleus that may be moving unpredictably slow would probably not know its whereabouts, while an electron that is going extremely fast would have no idea where it is. Since people are made of particles, a traffic cop stopped a driver that was speeding in Würzburg, Germany, who was a student of Quantum Mechanics and asked him, do you know how fast you were going? and the student answered "No, but I do know where I am." So, knowing key principles of Quantum Mechanics can be very helpful in the everyday life of particles and people.

- Quantum Mechanics and General Relativity are well accepted scientific theories that are mutually inclusive, the unifying truth lies within the expanding wave function of every spatiotemporal source at any point that reconciles the aspects of the two successful theories according to scale.

- Space and time are two sides of the same coin, and the wave function is the mint. Search for the source of the wave function and you shall find the kernel of growth of all there is, e^{-ct}.

- To beautify physics is to give it an object and a wave. It would be improbable for the wave function not to do what it is capable of doing.

- If a particle could speak, it would probably say to its wave "our new quantum wave theory may be unappreciated and undiscovered, but it is our theory." A theory is the embodiment of facts, and the facts speak for themselves.

- The uncertainty principle indicates our current inability to know exactly and simultaneously where all quantum things are and how they move through space only, not because things are indeterminate which is an overstatement.

- An observer does not measure the truth of nature, only the response of nature to the measurement. The response is the indeterminism of the measurement, not the indeterminism of nature.

- Clearly, the truth of nature was meant to be a probability or a possibility, not a measurement of certainty for the benefit of an observer. Causality relies on the truth of nature. The usefulness of causality in physical science depends on determinism, but absolute determinism of all phenomena is an overgoing. As long as causality and quantum mechanical probability are mutually inclusive, determinism has probably a good chance to be useful.

- The probability of the wavefunction is the common sense of its own reckoning. The difference between a probability and an improbability is up to the wavefunction.

- An emergent certainty of the wave function is a manifestation of its probability, and its three dimensions are possibility, reality, and truth.

- The wavefunction of Quantum Mechanics has a lot of decisions to make over time, the wavefunction decides what possible things become probable. However, do not hold the wavefunction accountable for love at first sight, that was your own decision!

- There is a wave nature to all there is. Nature has symmetry, all particles, matter, and space-time, have wave properties. Every particle or object moving through space has an associated wave. Is it all a matter of waves?

- Once you eliminate the improbable, whatever endures, no matter how unexpected or unpredictable, must be within the intangible six-dimensional spatiotemporal wave function.

- It is easy to believe that classical structures are typically constructed from the bottom up, why should reality not be the same at the quantum level? The six-dimensional spatiotemporal wave function is complex (real and imaginary). The wave function emerges from the interference of spatiotemporal waves to manifest gravitation among other things through its probability. Wave Theory, Curvature, Spatiotemporal Expansion or Contraction, and Gravitation, are classical manifestations of the wave function of Quantum Mechanics. Amplitude, frequency, period, wavelength, velocity, and phase are all attributes of the wave function which use the math properties of complex numbers including subtraction, during wave interference. This realization motivates a quantum theory of gravity such as "A Dynamic Theory of Space-Time: A Matter of Waves", a gravitational quantum wave theory.

- Rene Descartes' assertion that empty space-time was not really empty, was prescient. The fabric of spacetime is very malleable and it is the framework of all physical fields.

- The curvature of space-time produced by the interference of spatiotemporal waves is proportional to the local density of all

145

forms of energy, mass, torsion, expansion, compression, stress, and pressure.

- What A Dynamic Theory of Space-Time teaches us as a Quantum Theory is that everything we thought was relativistic is also quantum mechanical.

- Quantum Physics teaches us that things can simultaneously exist as a particle and a wave, regardless of size or mass. Then, everything in the world is complex.

§ 2. Science

- All the reliable truth and comprehension about nature is built upon the structure of logic of the divine design. Only natural systems, at any scale, are self-sustaining by divine design. Nature is not a thing; nature is a constituent of the divine creator.

- One way to know that a theory has been widely accepted is when anyone acts like a greater expert on the theory than the original theorist.

- To the greatness of time, nothing that happens is small change.

- There is sometimes a scientific impulse to theorize before one has empirical data. The researcher begins to imagine facts indifferently to suit a physical theory, instead of a physical theory to suit some empirical facts.

- If anyone thinks that he has seen farther than Galileo into the wonders of nature, it must be because in his mind's eye he was able to see farther than his telescope. What a loss not to be taught by a great teacher like Galileo di Vincenzo Bonaiuti de' Galilei!, but his brilliant work and ideas has helped to find understanding within oneself. Molte Grazie, Insegnante!

- The separation between physical reality and human concepts of nature is science. Curiosity and the expansion of knowledge drive basic research, while in pure research the understanding of the physical phenomenon takes precedence over the mathematics.

- Quantum Mechanics contains within itself the spatiotemporal topology of the Special or General Relativity. Topology is the property of spatiotemporal expansion or contraction, that does not change in Special or General Relativity, regardless of scale.

- The structure of space-time may change for an emergent spatiotemporal quantum wave that may expand or contract, but not the property of its spatiotemporal topology according to General Relativity.

- Unifying General Relativity and Quantum Mechanics in four dimensions proved futile, since Special or General Relativity has been a topology of the wave function of Quantum Mechanics from the beginning.

- The diameter of the Milky Way galaxy that contains our Solar System is approximately 1×10^{21} meters while the upper bound of the diameter of an electron is 2×10^{-18} meters. The topology of the wave property of the Special or General Relativity is as conformal from a single meter to the diameter of the Milky Way galaxy as it is from a single meter to the upper bound of the diameter of an electron.

- In physics or any other field of science, if you only focus on what has been done, you will miss what remains to be done.

- Absolute, and relativistic time, of themselves, and from their own nature flow equably with relation to their own wave function. Absolute space and absolute time are transcendental.

- The tempo of relations of space and time makes relativity emerge in the absoluteness of space-time.

- A quantum mechanical conundrum, "if anything can happen according to probability, there cannot be natural laws, consequently, an observer can only describe what is measured; however, since the truth of nature is uncertain to the measurement of the observer, then, there are natural laws."

147

- The equivalence of pressure and energy density is an observable law relating to natural phenomena. No pressure, no diamonds!

- Science is an incomplete view of what is observed. Faith is the acceptance of what is seen as divine. Science without faith is a work in progress.

- Facts are to science what principles are to faith. Science and faith are correlated, not reciprocal. Even a physicist needs a little faith in the theories of science.

- The purpose of science is not to disprove faith. The wisdom of faith is to welcome all discoveries of science. Science and faith are two sides of the same coin, the mint is enlightenment.

- There is nothing in faith that says you cannot find a physical truth with science; there is nothing in science that says you cannot find spiritual truth with faith. Either way, the truth will set you free.

- The nature of our minds was brought about by the same creative force that brought about the rest of nature. Hence, nature was intended to be self-aware, and to contemplate and examine itself. These qualities of nature may be the beginning of science.

- The facts of science are based on the observations of reality, and reality is based on things that consist of elementary particles that are fields of potential and waves of probabilities. An observer benefits from having faith on the facts.

§ 3. *Mathematics and Geometry*

- The natural mystery of the sequence of prime numbers follow the order of the spatiotemporal waves, not every mystery needs to remain a mystery forever, not even in mathematics.

- Not even zero is nothing. The existence and utilization of zero is something in itself.

- The search for physical constants in our universe does not stop our universe from constantly changing.

- Transcendental numbers abide by the power of algebraic methods, even though they arise when imaginary quantities are involved in the exponential natural growth.

- Physical laws and the geometry of complex numbers conserve their proportions if they are conformal through time. Where there is a wave, there is relativistic geometry.

- The universe has been more consistent so far about its laws of nature than the equations of science that have tried to explain it.

- Infinity and eternity may be visualized as fractals from a very narrow point of view. Infinity endows eternity which causes infinity to go beyond.

- Geometry is visible physical truth; mathematics is its beauty and acknowledgement.

§ 4. *Motivational and Inspirational*

- New ideas are meant to evolve, and history is meant to become the judge and jury of the original idea.

- If one brings to the world good and blessed knowledge of nature given by the creator, everything good one will learn about nature is based on something one already knows from the blessed minds of those that have come before.

- The awareness of ignorance is the beginning of true wisdom. The awareness of true wisdom comes when the spirit is born again.

- The act of thinking provides the human soul with the best ideas from the highest source of creativity.

- A child is a miracle of creation born to be the hope of the world because a child knows how to love. The love of a child turns into energy that can light up all the hearts in the world.

- Who would set a limit to our universe? Who would dare to declare when eternity ends? Only time would dare to set a limit to space, and only space would be there to affirm when eternity ends. Only the divine creator of all there is, who was, is, and will be, would resoundingly answer "I am. I am who I am."

- Our consciousness thinks of infinity outside of itself, but there is also the infinity within consciousness. Nothingness does not exist. Existence is infinite and eternal.

- Every probable universe in the infinite bulk of divine creation is a spontaneous event within the eternity of existence, and nature is an objective reality for a universe with life.

- To love is to know the merciful and kind word of the Almighty, to know the truth of all there is. Intuition, imagination, and knowledge are good, but to love is divine.

- According to the historical timeline of physics, 300 years ago, people believed that gravity was a force that pulled people and things to the earth, and gravity acted on objects at the speed of light, 200 years ago people believed that atoms were indivisible and identical with the same mass and properties for all elements, 100 years ago, people believed that the Special and the General Relativity were always correct, so Quantum Mechanics had to be wrong, and no large amount of energy could be released from an atom. Today, in the early 21st century, people believe that space-time is four-dimensional, General Relativity and Quantum Mechanics are incompatible, and there is no quantum theory of gravity. Clearly, it seems to be a fact that the only thing constant in what people believe about physics is change, so there is hope.

§ 5. Entertaining

- I have always suspected that being a physics writer would had been a career disadvantage. Hence, I would like to thank those who have never read my book for the final proof.

150

- Common sense, or the typical concept of physical reality, are not quantum mechanical. Clearly, Quantum Mechanics is for a different mindset.

- For Einstein, relativity was like courting a nice girl for an hour and it would seem like a second. According to an expert of Quantum Mechanics meeting nice girls may be uncertain. Once Paul Dirac asked Werner Heisenberg, why do you dance? Heisenberg replied that when there were nice girls he felt like dancing with them. After a few minutes, Dirac asked Heisenberg again, Heisenberg, how do you know beforehand that the girls are nice?

- There is a tale about someone asking a pathological liar, do you understand Quantum Mechanics? Well, you are the first person to ask me that question, but I know that there are only two people in the world who understand it. Who is the other one?

- Einstein said that the only real value is intuition. So, the only imaginary value is imagination. Only the imaginary value is more important than knowledge.

- Nature has no fear of perfection. Beauty has no regrets for being part of that perfection.

- Science is the process to understand the miracles and wonders of nature. If one loses one's faith in science, it will take nothing less than a miracle of science to get it back.

- Even to this day, above the old gates of the Temple of Science there is a sign that says, "Thou might not but hast faith."

References

Alcubierre, Miguel. (1994) *The warp drive: hyper-fast travel within general relativity*. Class. Quantum Grav. 11-5, L73-L77.

Baker, Bevan B., and Copson, E.T. (1987). *The Mathematical Theory of Huygens' Principle (Third edition)*. Chelsea Publishing Company, AMS, New York, NY.

Brown, Hugh Aunchincloss. (1967) *Cataclysms Of The Earth*. Twayne Publishers, Inc.

Caroll, Sean M. (2020) Interview on Events in Quantum Mechanics and Relativity with Robert Lawrence Kuhn on Closer to Truth, from a YouTube video dated November 12, 2020.

Einstein, Albert. (2003) *The Meaning of Relativity*, p. 113 (Psychology Press).

Einstein, Albert (1952). *Relativity, The Special and the General Theory*, Crown Publishers Inc., One Park Avenue, New York, NY 10016.

Einstein, Albert, Rosen, N. (1935). *"The Particle Problem in the General Theory of Relativity"*. Physical Review, Volume 48, July 1, 1935.

Eisberg, R. & Resnick, R. (1985). *Quantum Physics of Atoms, Molecules, Solids, Nuclei, and Particles* (2nd ed.). John Wiley & Sons. pp. 59–60. ISBN 978-0-471-87373-0.

Everett III, Hugh. (1973) *The Many-Worlds Interpretation of Quantum Mechanics*, Princeton Series in Physics, Princeton University Press.

Fuller, Robert W. and Wheeler, John A. (1962) *Causality and Multiply Connected Space-Time*. Phys. Rev. 128, 919 – Published 15 October 1962.

Gaspar, Enrique. (1881) *El Anacronópete*. The Time Traveler, published in 1887 in Barcelona, Spain, by Daniel Cortezo y Compañia. The first known drawing of a time machine.

Hapgood, Charles Hutchins. (1958) *Earth's Shifting Crust: A Key to Some Basic Problems of Earth Science*. Pantheon Books Inc. (Foreword by Albert Einstein)

Hey, T., and Walters, P. (2009) *The New Quantum Universe*, Cambridge University Press.

Huygens, Christiaan. (1690) *Traité de la Lumière*, Leiden: Pieter van der Aa. (translated by Silvanus P. Thompson, 1912) Treatise on Light, London: Macmillan.

Melia, Fulvio. (2007). *The Galactic Supermassive Black Hole.* Princeton University Press, 41 William Street, Princeton, New Jersey 08540.

Nieves, Robert. (2020) *A Dynamic Theory of Space-Time: A Matter of Waves*. Published by Kindle Direct Publishing, Amazon.com, Inc. ISBN 9798667276289.

Rucker, Rudolf v. B (1977). *Geometry, Relativity and the Fourth Dimension*, Dover Publications, Inc., New York, NY 10014.

Schwerdtfeger, Peter, et Al. (2008) *Relativistic and electron correlation effects in static dipole polarizabilities for the group-14 elements from carbon to element Z= 114: Theory and experiment,* Physical Review A 78, 052506 2008.

St. Fleur, Nicholas (2016). *Four New Names Officially Added to the Periodic Table of Elements*. New York Times. (December 1, 2016).

Staff (2016). *IUPAC Announces the Names of the Elements 113, 115, 117, and 118*. IUPAC. (November 1, 2016).

Subramanian, S. (2019). *Making New Elements Doesn't Pay. Just Ask This Berkeley Scientist*. Bloomberg Businessweek.

Taylor, Edwin F, Wheeler, John Archibald (1966). *Spacetime Physics*, W.H. Freeman and Company, 41 Madison Ave. E 26th, New York, NY 10010.

Wald, Robert M (1977). *Space, Time and Gravity, The Theory of the Big Bang and Black Holes*, The University of Chicago Press, Chicago 60637.

White, George H. (1978) *Time Travelers, The Great Saga of the Aznar*. Volume 16. Eurocon, Brussels, 1978. Published by Silente Ciencia Ficción.